Living on an Acre

2nd edition

A Practical Guide to the Self-Reliant Life

U.S. Department of Agriculture
Fully edited and updated by
Christine Woodside

LYONS PRESS
Guilford, Connecticut
An imprint of Globe Pequot Press

To buy books in quantity for corporate use
or incentives, call **(800) 962–0973**
or e-mail **premiums@GlobePequot.com.**

Lyons Press is an imprint of Globe Pequot Press.

Library of Congress Cataloging-in-Publication Data is available on file.

ISBN 978-1-59921-885-4

Printed in the United States of America

10 9 8 7 6 5 4 3 2 1

Contents

Foreword to the Original Edition, *Living on a Few Acres*

The New Back-to-the-Land Movement

Americans keep going back to the land. It is a pilgrimage that makes more sense to a lot of people than living in cities, enticing us with the promise of escape from freeways, assembly lines, and crowds.

The land offers freedom, a chance to test your mettle against nature's challenges.

The pilgrimages began with our agriculturally-minded, freedom-loving forebears of the seventeenth and eighteenth centuries. In waves since then—from the founding of community utopias in the first half of the nineteenth century, to the homesteads of the later 1800s, to the flight of the unemployed from a collapsing economy of the 1930s—Americans have returned to the land.

Now the tide of Americans that swept to the cities after World War II has ebbed. The flow has reversed. People are populating the countryside faster than they are cities.

I know their motives. I am a farmer.

So I know that country life can push some people beyond their endurance, can shatter illusions with a heavy and indifferent hand, and can press poverty upon the backs of the unlucky and the unprepared.

This book is intended as a practical guide for those who make the journey back to the countryside and for some of you who are already there. It is mainly for those who intend not to gain their principal income from the land, but rather to have a job in town or live on a pension or some other source of income.

USDA programs offer valuable assistance at the local level. In addition, this department has a major responsibility for all federal efforts in rural areas.

Living on a Few Acres describes both the pitfalls and the satisfactions of country life. There are plenty of both. And there is nothing quite like country living.

—Bob Bergland, former Secretary of Agriculture, 1978

Editor's Note: Homesteading fever ended for a while in the 1980s, but the numbers swelled again in the 1990s and interest has not flagged since. Half of all American farms are small-scale operations, with income of anywhere between $1,000 and $250,000 a year, according to the USDA. Most rural areas in the United States are growing steadily and quickly. Demographers find the strongest flight to quieter lands is taking place in the Mountain West, the Upper Great Lakes, the Ozarks, parts of the South, and the rural Northeast. Rural populations are declining in the Great Plains, Western Corn Belt, and Mississippi Delta.

Editor's Preface to the Second Lyons Press Edition

This book helps people go forward on their land by going back in time. It is a revision of a classic book that sat on homesteaders' shelves for a generation or two, but it includes twenty-first-century knowledge, technology, and perspective. Meet the smart USDA staff that tried to help the hippies avoid disaster when they bought their tracts of rural land during the homesteading boom of the late 1970s.

Buoyed by energy shortages, the environmental movement, and a recession, homesteading fever gripped many Americans, just as today's local food and farmers' market movements seem an antidote to energy, climate, and financial pressures. The United States Department of Agriculture, perhaps sensing that new enthusiasts might not know what they were getting into, decided to devote its yearbook of agriculture for 1978 to the movement. They called the book *Living on a Few Acres*. The USDA's yearbooks of agriculture compiled the best advice for anyone who desired to learn from 1894 through 1992, when, unfortunately, they ceased publication. The yearbooks even now make a distinguished set of themed guidebooks for farmers, would-be farmers, and anyone who wants to understand soil, climate, growing, and raising.

The 1978 volume followed that pattern by laying out practical, general advice from experts who talked almost as if they had stopped by your house to chat. *Living on a Few Acres*, edited by Jack Hayes, addressed people who wanted to try farming on a small scale and experienced farmers scaling back to part-time work for retirement. The book you are holding is an updated version of it. I have worked hard to leave intact the blunt yet sympathetic tone of that book while adding up-to-date data and sources to help you. The 1970s were such an enthusiastic time. They were also a slightly nutty time. Everyone had a dream that could be found embroidered on a decorative pillow.

The author of the foreword was then-Secretary of Agriculture Bob Bergland. His advice to test yourself and enjoy the freedom of making a living from the land holds up today. He understood that this could be very difficult. He was himself a farmer.

A. Gene Nelson co-authored the important first two chapters about what it's like to live in the country while considering the tradeoffs of leaving the city. At the time, he was an extension farm management specialist at Oregon State University. Since 1990 he has been a professor of agricultural economics at Texas A & M University. I was able to get in touch with him, and he wrote: "Now that the USDA has discontinued publishing these yearbooks, you are providing a service by updating them. Much of the information is probably still relevant."

We agree. Although we renamed this book *Living on an Acre* to reflect the shrinking size of country tracts, it serves the same purpose as the original. This is a practical guide to several enterprises. It is mainly for those who have not tried this experiment before, most of whom will keep their jobs in town (or in a home office, perhaps) or will live off pensions or investments. Ninety percent of farm income in the United States comes from off-farm employment. If only it weren't true, but that is the way it is. Really small farms, fewer than 10 acres, make up about 10 percent of all farms in America. Think of that—an important minority of a vital industry supports itself on capital earned off the farm. So you aren't too wet behind the ears. Read on, and step back into the hopeful 1970s homesteading movement. Apply that hope to your land now.

—Christine Woodside, Deep River, Connecticut, November 2009

Acknowledgments

The following people provided invaluable help in bringing the information in this book up to date:

Debbie Smith of the United States Department of Agriculture's communications office; Nancy Weiss of Pomfret, Connecticut, and Arthur Johnson of Darlington, Maryland, who wrote new essays about their farming experiences of the last twenty-five years; Robert Ricard of the Connecticut Department of Environmental Protection Division of Forestry, for information on trees and windbreaks; Dwain and Mary Bartel of West Bend, Wisconsin, and Arthur Johnson of Darlington, Maryland, who discussed their experiences with dairy and beef cows; Scott Konrad of Essex, Connecticut, for insurance information; and Duo Dickinson of Madison, Connecticut, for green building information.

PART 1

Pluses and Minuses

Living in the Country— a Diversity of People

Whatever your vision of country people might be, rural residents are a diverse group—computer analysts, chiropractors, carpenters, salesmen, professors. Their land ranges in size from enough for a rural house with a large backyard to small-scale farms involving a number of acres. The locations stretch from Vermont's rocky soils to the Blue Ridge foothills of Virginia, from the prairies of Iowa to the southlands, the western deserts, and the Northwest's lush forests.

There is no such thing as the average country resident. People live in the country and own an acre to several acres for a variety of reasons.

- The rural residents use their land solely as a residence.
- The hobby farmers work the land for recreation.
- The gardeners provide some of their own food to save money.
- The alternative farmers provide an alternative lifestyle for their food and energy.
- The part-time farmers aim to supplement their income.

THE RURAL RESIDENTS. These people sell less than $1,000 of agricultural produce a year. Their primary interest is the pastoral setting for their home. Many view the few acres on which they live as an extended backyard. They are often city folks buying up old farmhouses, plus 5 acres to keep a horse. In other cases they are retired farmers who might be renting out their farm while continuing to live in the house. Sometimes rural residents tried farming for a while, then decided to give up and get jobs in town, but to keep living in the country.

Two couples formed a partnership to farm land that included Christmas trees, woods, and pasture. They didn't have any experience with this life, but they wanted to live in the country, and banding together made it happen.

THE HOBBY FARMERS. These farmers have income from their land as a secondary goal, but work a few acres to pursue their recreation and hobby interests. For some it might include woods and a stream for hunting and fishing. Others may have horticultural hobbies and devote their leisure time to landscaping. A modern "gentleman farmer" with the right resources can hire help to maintain a pretty nice farm, especially when there is no pressure to make it profitable.

THE GARDENERS. They want mostly to improve their quality of life and, along the way, reduce their family's food bills, by growing their own food. The local food movement has gained new strength since 2000 or so, as many people understand the health and environmental benefits of local produce that is fresh and did not have to come to your table by airplane or long truck trip. These gardeners may use their land for an extensive vegetable garden, one or two head of livestock, and some fruit trees. They put in a lot of time because to them, time is not money. The seasonal farmers' markets are full of these devoted planters, who truck in their fruits, vegetables, maple syrup, honey, and herbs for a day at a time. "We save a considerable amount of money on vegetables—especially tomatoes—if you don't figure in the time I spend canning them," said one small-scale farmer.

THE ALTERNATIVE FARMERS. These rural dwellers seek to minimize their reliance on services, particularly electricity or gas lines. They tend to grow organic food and they might use animals to farm. They do not use chemicals, and they avoid pesticides.

THE PART-TIME FARMERS. They have definite income goals for their limited acreage. They are seeking through use of their land, labor, and financial resources to earn a significant portion of their income. They may be retirees supplementing their Social Security or pension income, or they may be seeking to augment their salaries or wages earned off the farm.

Be Realistic

Success on an acre can mean you found a peaceful home, a good investment in land, or a cash income. The cost of living in the country usually is lower than in town, especially if you raise some of your food and don't have to spend much on transportation.

Achieving this life can be challenging. If you are from California and move to southern Oregon, determined to grow peaches and walnuts, for example, you might learn too late that the cold winters will kill these crops in which you invested your savings. If you have never bought country land before, you might buy the wrong kind of land and discover that only a fraction of it will grow anything. If you will be keeping your job, you might learn the hard way that you're too tired after work to garden or tend animals.

A primary difficulty stems from the romantic (that is, uninformed) notion of rural life. Many people know nothing about animal and plant diseases, insect pests, and other production and marketing problems they may encounter. Or they ignore them. As a result, they often find trouble. An almost certain way to invite disaster is to buy land before developing a realistic plan for what to do with it. Make sure this doesn't happen to you, even if you can sell the land without losing your investment. Take the attitude that you are going into your rural life prepared, with the odds in your favor.

The ultimate success of a farm operation depends on how well you can cope with challenges, and how strongly you feel the advantages of rural life outweigh the disadvantages. Problems result from changes— changes in weather, or in your own goals, values, or resources. One of the changes that occurs over time is the composition of the rural family. Children are born, grow older, and eventually leave home.

The happiest seeker of the rural life may be the part-time farmer who commutes to his city job and feeds his family and his soul with a large garden and a few animals. But people living on an acre all have their individual methods of making it work.

Consider the Trade-Offs Before Leaving the City

Noise, congestion, traffic, pollution, a lack of privacy, high crime rates, concrete, and other things about the city are finally beginning to get to you. Quiet, clean air, pets, homegrown produce, and open space all sound very appealing. Of course, neither city life nor country life is perfect. Choosing one is a matter of making trade-offs. If you pick the country, you must believe deeply that the values and the satisfactions of rural life really outweigh its inconveniences.

Making a permanent move from city to country for a whole family can require a big adjustment. Families have made the move and have been quite happy. Some wish they hadn't, or the children grow up and they go back to the city. (Remember that children do grow up, and the family's values change over time.)

Spending weekends or a summer vacation in the country with good friends or relatives may well be remembered as heaven on earth, but was it the change from the daily routine that made the experience so delightful, or was it truly the advantages of country living?

It was fun to care for the livestock in the spring, but livestock have to be cared for every day of the year. Getting out of bed on a warm spring day is far different from getting out of a cozy bed to go out in the wet and cold to care for the same animals.

You may never be happier than when you're working in your small garden in the backyard in town after a hard day in the office, and perhaps it would be nice to have a larger space so that more produce could be raised. But will it be that much fun after the longer drive home from work and with a much larger task facing you?

Sure, country living is great, but there's no denying that some people should never move out of the city. They need the things a city offers such as convenience, entertainment, shops, nearness to people, and medical care.

Before making a decision to move or stay, you might draw up some sort of a balance sheet listing the pros and cons of what your present life offers compared with what you feel a move to the country would offer.

Do not believe that the country insulates you from environmental problems. Air pollution may be less or different in the country, but it is not eliminated. You may be trading industrial smoke and car exhaust fumes for animal waste odors, crop and weed pollens. Dust storms can be a problem in some areas. Particulate air pollution, called haze, hangs in the air of many rural areas of the East Coast. Winds carry it in from industry as far off as Ohio.

As a rule, the country offers more opportunities for outdoor living than the city. Your house is seldom close enough to the next one to make screens and fences necessary for privacy. There is usually room for a good-sized lawn for recreation. But remember that a lawn is going to take more time and care than the small city plot, thus leaving less time for the fun you crave.

Most places in the country are closer to hunting and fishing than are the cities. But consider that the novelty of doing these things might wear off, or that your family might not like the idea that instead of hunting every weekend, you now want to do it every day.

Your new next-door neighbor may live $1/4$ mile away. After your spouse heads off to work, and the children to school, it may be more difficult to find someone to meet for coffee. There might not be time for coffee in any case, with so many chores around the place.

Opportunities for New Technologies

Are you moving to the country to save energy? Unless you are committed to that goal, you might not. Data on home energy use shows little difference between a country house and a city dwelling. Per capita, early-twenty-first-century Americans use more electricity and oil to do basic chores than they need to.

However, in the country, your chance has come to change that. You will have the room to use alternative energy sources such as photovoltaic solar collectors and even wind turbines, which save much of what you'd pay for power off the regular grid. Seek out information on these technologies and realize that the savings come after a few years.

The food you raise yourself may be better and less expensive than that available in the city, but it may not be. Don't expect all your fruits and vegetables to look like the beautiful catalog pictures. The directions printed on seed packets are helpful but incomplete. They don't say what to do when you see little green bugs or when the leaves turn brown.

On the other hand, you'll never purchase sweet corn in the city market that compares with that you've picked yourself at the peak of perfection, and have on the dinner table twenty minutes later. Try to be realistic about how large a garden you can handle. If it gets to be too big, you will have to buy a tractor and a complement of machinery. You probably won't be able to eat all the produce when it's fresh. If you don't plan to sell any, you'll want to can or freeze it.

Maybe you have always dreamed of having a little more space so you could have some animals. It's a great experience for children to learn to care for and love animals, but animals need care every day. Chickens don't require much space, and neither do sheep and goats, but horses and cows do. Raising part of the feed requires more acres and time, leaving less time to enjoy the animals. Buying all the feed gives more time for the animals but less money to spend for other purposes.

If you plan to raise some animals for meat, buy all the feed, slaughter the animals, and process the meat, it is unlikely that you'll end up saving much money. Some of the food you raise yourself may cost more than if you purchased it in the store. It's hard to compete with the commercial egg producer who has 100,000 birds, automatic equipment, and the management to cut costs and maintain quality. Remember, too, that eating pork chops from the pet hog that grew up in the backyard can make an awkward situation for some at the dinner table.

Often the hobby gets a little expensive, so people decide to expand in an effort to supplement the family income. In some cases people with small plots have done quite well financially on a part-time basis, but they have some things going for them that newcomers will not have. Most have good farming backgrounds and experience. They really enjoy their work and are willing to put in several hours of work each day or week, and usually they concentrate on some highly specialized crop adapted to top-quality soil and an ideal climate.

Some small-scale farmers have been encouraged to raise specialty crops by the promise of a good market. Too often and too late, they have not found that market adequate. Starting a new venture on your own means that no one will train you, as is the case when you are working for someone else. True, much good information is available, but when you make a mistake, there's no one else to pay for it. You become the risk taker.

While you can make money in part-time farming, there is another side to the coin. You can lose money, too.

Conveniences

Living in the country, for all the peace and quiet, has drawbacks. You may find some difficult to live with, impossible to change, and you may miss the advantages an urban area offers. Consider a few things you take for granted in urban life but can't find in the country (except at a high cost).

In the city a reliable supply of water flows from a faucet. Water usually flows from a faucet in the country, but the source of the water is more apt to be your own well and pump. When the pump fails, it is your responsibility to get it repaired. When the power goes off, the pump does, too. Be prepared for that inevitability with a backup pump.

Most rural residents use septic tanks that generally work well if slope and soil conditions are right, but when things go wrong, it's your responsibility to solve the problem. In the city, collectors take away your garbage, but in the country, you do—or you have to pay someone who will.

Consider the time you will need to get places. Schools in rural areas may be far from home. Children sometimes meet the school bus before daylight and get home after dark. They might not have the same courses or extracurricular activities they had in larger districts. Parents must get them to and from activities, which can mean a lot of time in the car.

In the country, the neighborhood store may be several miles from home. You will have to change your shopping habits or you will expend many miles, a lot of gas, and a lot of time.

While in the city the challenges of driving are dealing with traffic and finding parking spaces, in the country the problems are the roads themselves. Snow or mud might close them. Learn to keep a stock of essentials in case you are stuck at home for a day or two.

The nearest fire department, ambulance station, police station, or sheriff's office may be miles away. Long distances from places like these will be reflected in higher insurance rates. If roads are poor or the area is isolated, it will take a long time for emergency help to arrive. Remember, taxes may be lower in the country, but so is the quality of the services.

Adjusting to Changes

The change from city to rural living often creates severe adjustment difficulties—loneliness, learning to care for equipment, dealing with animals all the time. None of these problems are insurmountable, but certainly should be considered when listing the pros and cons of city and country living.

Those who stay at the country home during the day might feel left behind. Some people who move to the country find new financial stress. Equipment to maintain the place becomes essential since services are so isolated. Repairs to animal fences, and upkeep of buildings and equipment, suddenly take part of the monthly income. Some of the new experiences often become daily or weekly aggravations.

Chances are, if you were happy where you came from and your family has the ability to adjust, you will be happy in the country. Country living can be great, but it may not be as cheap or convenient as you thought.

So remember, it's relatively simple to come and go when you live in the city, but in the country your freedom can actually tie you down. If you own animals, you must find someone to care for them. You have to time vacations so that they don't coincide with the planting, caring for, or harvesting of your crops. Sudden and severe storms in the winter can damage properties left unattended.

You'll have some wonderful experiences even after considering the points discussed here—and others you may think of. You and your family are the only ones who can determine what you are willing to tolerate on your way to getting the life you want. No one else can determine this.

Face the Realities

Are you ready to deal with the realities of getting started? Some questions, and answers, about your goals and resources will help you decide how to enter small-scale or part-time farming.

- Do you want to be a part-time farmer on a few acres and keep a full-time job in town?
- Can you find a farm that suits your needs in a place where you are willing to live?
- How much will it cost? If more than your savings, can you obtain financing?
- Do you have the knowledge and skills needed to operate a small farm?
- What about financing the production process?
- How and where will you sell your produce?

Part-time farming is usually the choice of people who want to get out of town, but not too far, and to have a little room, but not too much. Space will be available for a few fruit trees, a vegetable garden, even some poultry and animals to reduce the family food bills. A crop can be produced that will earn part of the needed family income.

Small-scale farming on a full-time basis is the goal of some urban families. Fed up with living and working in what they view as crowded, noisy, impersonal cities, they seek the more tranquil life and the satisfactory financial rewards they believe await them on a small farm.

Others will choose a compromise between those two extremes. Their goal is full-time farming, but they accept the reality that it may not be within reach. One or more family members will have to work elsewhere to supplement the income earned by farming a few acres of land. They can enjoy the advantages they associate with rural family life, which will be diluted by the necessity for some to commute to work, perhaps for long distances.

Some people truly want to get away from civilization, at least for a while. They are willing to live very modestly, create with their hands, and produce with their own efforts to provide for their needs. Shunning many of the offerings of modern civilization, they choose to rely heavily on land, livestock, native materials, and hard work to accomplish their objectives.

Your goal will be more easily reached if all family members who are old enough to participate have helped to make the decisions. Along with the many benefits you will enjoy, much hard work lies ahead. A family united in purpose and working in harmony to reach the agreed-upon objectives will more easily deal with problems, frustrations, and adjustments.

Your Resources

If you are typical, you probably have three main resources for making your land produce: *labor* (including yours), *management* (probably you), and *capital* invested in the land, equipment, and operating expenses. These resources will dictate the type of small farming enterprise at which you can succeed.

You are going into competition with commercial agriculture and must face some of the realities of this competition. Most of America's agricultural production is highly mechanized. In the growing of many crops, a small unit cannot hope to compete with a typical large-scale farm. But this is not the case with labor-intensive crops—that is, those that require many times the number of hours of labor per acre needed by crops that are almost completely mechanized.

Labor provided by family members is a major source of income from small-scale farming. If you are going to sell your labor effectively on a small unit, you may want to select from crops such as fruit trees, grapes, vegetables, and others that require intensive use of labor, perhaps more than 150 hours per acre each year.

You can compete if you are willing to do the extra work. It takes just as long to hand-prune a tree in a 5-acre orchard as it does in an 80-acre planting. By contrast, producing barley on a large-scale farm requires much machinery and only one to three hours of labor per acre each year. There is little opportunity here to earn labor income unless a large area is farmed.

So, if you are energetic, have plenty of help available, have production expertise, and all goes well, the right choice of crops will provide an outlet for all of this energy at a reasonable return. Labor and all the other production inputs must be used at the right time and in the right way or there will be neither profits nor returns for your labor, management, or capital. Your ability as a manager of your resources will determine whether you receive a return on your investment and for your labor. There is no guarantee here; it's up to you.

Getting Located

Decide both *where* you will farm and *how* at the same time, but be flexible. The larger the general area that suits your needs, the easier it will be to find your farm. Your choice of location and of farm enterprises may conflict. Weigh both carefully while you make your plans.

You are investing in both a small business and a home, so this is the time to decide where you might want to live for many years to come. Consider all areas of interest and types of farming that appeal to you before you make a decision. It is a good idea to travel through these areas and to spend some time exploring them.

Zoning is usually a fact in the country as well as in most cities. Generally speaking, areas next to cities are zoned for small acreages, perhaps 1 or 2 acres or as many as 5.

You can learn a lot about areas and people by subscribing to community newspapers and by becoming acquainted with local business and professional people. To learn about the usual high and low temperatures, annual rainfall and similar data, contact your nearest U.S. Weather Bureau office. Cooperative Extension personnel can be helpful in providing information about types of agriculture in the areas you visit.

The more time you take to look around, to meet and talk to people, and to accumulate information, the more certain you will be that you picked the area best for you.

More for Less

Have you already looked at real estate advertisements? Land can be expensive in areas where country retreats are in high demand.

Does the place you are interested in have a house on it? If so, is it one you would be willing to live in during a cold, wet, windy, disagreeable winter? How about a hot, dry, windy, dust-blown summer? Building or remodeling is expensive, unless you are capable of doing most of the work yourself. Labor will be about half the total cost for most reconstruction.

No purpose would be served by attempting to quote prices here, since each unit will be priced by its owner, and these prices change rapidly.

However, small acreages of suitable farmland close to growing urban areas generally cost far above any value represented by income that can be earned from farming.

You will need some machinery, perhaps an irrigation system, and some sheds or storage buildings. If you plan to keep livestock, you must provide fencing and shelter. People are often surprised by the high cost of farming—including mortgage interest payments, equipment purchases, and seasonal crop expenses. Contrary to popular belief, land will not always produce enough to pay for itself.

Land and Water

Is the quality of the land suitable for your farming project? If it is presently being farmed and crop production is satisfactory, it is unlikely that any problems exist. But someone may have planted tree or vine crops on land not suited to their production and you could be buying trouble. If you are looking at open land, have some soil tests done, and get an expert opinion to satisfy yourself that your planned crop production could be successful.

You will certainly need water for domestic purposes and for livestock. In many areas of the country, water will be needed to irrigate your crops. Some areas have adequate supplies that are suitable for these purposes. In others, water is limited in both quality and quantity.

If you are planning to buy open land and drill a well, get information about the groundwater supply. Check on the likelihood of having enough irrigation water if more land around you is developed. It is a good idea to get professional advice and also to check with neighboring landowners who are irrigating crops. Add their judgment to other information you gather.

Will you have the right to drill a well or to use water from a stream? Is a permit required? Ownership of water rights is a complex matter. More than just a few times the new owners of property bordering on a beautiful mountain stream have found, too late, that they had no rights to any of the water.

Laws concerning water rights differ from one state to another. Prospective buyers will do well to find out ahead of time whether any problems about water rights will concern them. If surface water is involved, check also to see if the babbling brook that is so lovely in the late spring is still flowing at the end of a hot dry summer.

What about labor? If you plan to develop or grow crops that demand extra labor at certain times of the year, can you find that needed help? Casual labor may be available from a nearby school, community, or college. You may need to make arrangements well ahead of time and perhaps furnish some transportation or temporary housing to attract the extra help.

Are you counting on family labor to get most of the work done? This can be the answer, but on the other hand, after the novelty wears off, some family members may find they don't like farm work and may choose to sell their marketable skills where the hourly earnings are higher than on the family farm.

Even though the climate may be generally suitable for the farming program you would like to follow, there can be localized or spot problems. Check weather records for the possibility of a late-spring or early-fall frost if you are going to raise crops that could be damaged. Are there any peaks of high temperature that will normally occur and cause problems? If you are planning to raise crops that are not produced nearby, be especially alert to climate-related problems.

Do you have a marketing program in mind? Some crops are handled by marketing cooperatives, while the producer must sell others. Sometimes a roadside market will provide an outlet. Local retail food stores may be interested in what you plan to grow. Check on all possibilities and be sure you have available markets. You will need this information when you are arranging the necessary financing to buy your farm as well as when you have products to sell.

Knowledge and Skills

Do you have the knowledge and skills needed to operate the types of small farms that interest you? If you are counting on a reasonable amount of farm income, then you must have or acquire the ability to manage and operate this resource.

Farming skills can be acquired in many different ways. If you do not have them now, you have many sources of help available. Ways to acquire needed information include Cooperative Extension programs, agricultural courses in high schools and in community colleges, and working on farms. Many of the crops most suited to small-scale farming require a great deal of production knowledge and skills. Applications of chemicals to the ground and perhaps to the crops must be done at the right time, in the right way, or total crop failure could result. A few days' error in timing can also create disastrous results by ruining the quality of the crops.

Perhaps you do not want to use any chemicals for weed or pest control; maybe you do not want to use fertilizer either. Most crops must have some protection from insect pests if good yields of high-quality products are expected. You must learn alternative methods of pest control.

Compost and animal waste may be your choices for fertilizer. Knowing how to make compost and where to locate organic fertilizer are skills and information you must have. You are going to own a modest amount of farm machinery as well as the usual number of home appliances, and you may live a long distance from repair shops. The cost of bringing repairmen to your small farm will be far more than you were used to paying in the city. A money-saving idea would be to learn about machinery repair and household maintenance.

Money Matters

Consider your financial resources. You may need to borrow money to operate the farm in addition to buying it. This may influence your decision between buying an established farm or buying open land and developing your own enterprise.

It is usually easier to finance an established farm than to finance open land, housing, and crop-developing costs. If you decide to buy land and plant permanent crops such as trees or vines, it will take about three to six years of expenses before you have very much income. Your financial resources may carry you, but if not, someone in the family has to get an outside job.

Before you make an offer on land or a farm, you must estimate the cost of improvements, labor, and the cost of the farm operation. Plot a month-by-month schedule of the cash cost of production for each crop or livestock enterprise. Also, plot a total cash-flow budget for your country place.

Start with the cash cost of production schedules. Charge for everything you will buy and use. Project the labor required during each month. The total for each month can then be measured against the amount you and your family can accomplish. The labor schedule will help you plan for the times during which you must hire help or arrange to trade work with neighbors.

The cash-flow budget will show accumulated total farm cash costs and income received during the crop and livestock production cycles. These budgets should include an amount for family living and an operation loan, if one is required.

You should be able to take out short-term loans for operating expenses. Include your income from an outside job in the cash-flow budget. This will show how much you have available to pay off the debt.

There are many reasons why you and your family may choose life on a small farm unit. These include the complexities of urban living, a need for greater freedom and independence, and a desire to be more in control of your own lives. There is no one formula for success in meeting these expectations. Still, an energetic family working together on suitable land with a good plan to follow, a willingness to work hard, and the skills to manage all of its resources has the greatest opportunity to achieve its goals.

Changing to a New Lifestyle: Little Things Add Up

You probably have thought a lot about the transition from settled urban life to rural. Still, moving to the country can be a shock. Most people welcome the change, but many find the transition too much of a sacrifice. A number of people have become disgruntled, viewed the initial decision to move as a big mistake, sold their error at a loss, and moved back to an environment similar to the one they left. This is partly because there are certain amenities in everyone's way of life that don't appear to be very important—until you give them up.

Take the time to sit down and make some lists. Evaluate the things that make you happy, and the value of things you must give up or tolerate in a different environment. Many of the activities that are part of the urban community are not as accessible or abundant in rural areas. In the country, there are fewer theaters, restaurants, stores, repair shops, doctors, hospitals, etc. Some people get a great deal of satisfaction out of shopping and comparing from one store to another. In many rural areas you usually won't have as many options.

Tranquility or Boredom?

When you first buy the place, it is a delight to visit. Indeed, there may be an abundance of things to do in the little time you have to spend on those long weekends and vacations. But you also find time to sit back, relax, and enjoy the peaceful serenity of your few acres and the sweet country air.

If you make the transition to permanent residence, you may find yourself catching up on all the improvements and renovations that were planned. The lifestyle in general is at a slower pace than in urban areas.

What once appeared to be peace and tranquility can become boredom and dissatisfaction for some people. This is not to imply that there is nothing to do in the country, but that the choices may not be particularly satisfying for some on a routine basis.

Caring for livestock, for example, is a reasonably minor chore in good weather. When the cold hard winter comes, however, and pipes freeze and burst and you have to go chop ice so the livestock can get water twice a day regardless of wind, rain, sleet, or snow, that previously enjoyable activity becomes a job. It's a good job if you like it. But if you're not sure, find out and seriously think about it before you make the commitment.

Other important considerations that can make a difference in how happy you are with rural life include the availability of off-farm employment opportunities, neighbors, friends, playmates for children, and the family's consent on making the change. There will be fewer neighbors; friends will be farther away, as will be playmates for children. One unhappy member of the family can make all the rest uncertain about the new lifestyle. Getting back and forth to social, civic, and athletic activities will require planning, coordination, time, and sacrifice by some of the family members.

Balance the family's desire for a rural life against the need to get to town for work and social life and the drawbacks of your new life. You must adjust seasonal and daily patterns of life. The decision to raise a small garden or a few chickens is one thing, but the keeping of milk cows or dairy goats—which must be milked twice a day about 300 days a year—is something entirely different. All the pleasures and enjoyment of country life can be quickly lost when faced with a never-ending cycle of chores. Poor choices of enterprises or combinations of enterprises can lead to trouble and frustration.

Therefore, to help avoid the danger of "getting in too deep," or the feeling that you're tied down to the place, be aware of what each new undertaking requires. It's better to start a little slow in developing your land than to rush into something unprepared or unsuspecting.

By carefully planning, coordinating, and controlling the various farmstead enterprises and activities, rural living can provide a healthy, wholesome mixture of productive work and recreational opportunities.

Realistic Goals

No family should get its hopes too high in any new venture, and this is particularly true of adjusting to the rural scene on a full-time basis. While beginning stages of the transition can inspire you to try to be as self-sufficient as possible, the fact remains that the cost is just too high. Although making your own butter, bread, and preserving most of your foods may sound gallant, healthful, and natural, you must remember that only the experience of many years can make your dream come true.

Limiting the amount of necessary work, and adding more variety to your country place gradually, can ease the transition and prevent overwhelming frustration. When you can handle a small garden with a few chickens, for instance, you may then consider the addition of a family milk goat, or even a cow. However, for the sport-minded, happiness may be a horse or pony instead, or for the specialized hobbyist, a few sheep for wool to process, or a rabbit project for furs to work and sell. It's up to you, but proceed with caution.

Getting away from the routine of rural life may be necessary from time to time, and vacations should be part of your established priorities. While it may be more difficult to get away than for urban dwellers, it is not impossible. With a little planning and preparation, various arrangements can be made.

For example, most people who relocate to a rural area still have urban friends who enjoy regular visits to observe the rural scene. They might dream of a future venture of their own. These families are usually more than happy to take over your feeding chores when given specific, written instructions, and emergency phone numbers, such as your veterinarian, an experienced farmer, and where you can be reached if necessary. While you may see this as an imposition, most people consider it a welcome learning experience for their children and themselves.

Other choices of arrangement might include paying a teenager from a nearby farmstead, or swapping chores during vacation with another family with a similar setup. An extended trip may mean using a combination of several people to help at different times.

Whether you attempt to become totally self-sufficient and live entirely off your few acres, or decide to just raise the family's fruits and vegetables in a backyard garden, be prepared for certain facts of rural life: Water freezes in winter—crops fail—fences fall down—and livestock get sick. Realizing that sickness, injury, and disappointing harvests can occur, and being able to cope with these events, are all part of the rural experience.

So even though the best-laid plans don't always pan out, more often than not it is an educational experience that provides an opportunity to understand your environment and to constantly learn new ways to use nature's resources to your advantage.

Will You Be Happy?

This is not an easily answered question, and, in any event, only you can answer it. If you are honest with yourself and know yourself well enough, you can find out a lot by taking a personal inventory. An appreciation and tolerance of nature, of course, is essential. The quiet of the countryside, the smell of freshly mown hay, the sight of livestock grazing in a lush green pasture, and the taste of vine-ripened tomatoes are there to be enjoyed. However, nature can also be harsh and unpredictable.

Being resourceful is important. If you have to call in outside help whenever a fan belt in the car breaks or the lawn mower needs a new spark plug, you will spend a large amount of time and money at the repair shop. Minor problems, such as when Japanese beetles infest the vineyard or the calves get scours, are less bothersome when you can solve them yourself.

Patience is a must. Life in the country is slower paced. City dwellers seldom build or grow anything at the same scale as you will in the rural area. It takes time and planning to plant a garden or build even a modest-sized greenhouse. In the garden it takes from four to six months of careful nurturing before you can taste the fruits of your labor. In the orchard it may take four to six years before the first apple harvest from those newly planted saplings. An appreciation for and acceptance of nature's time schedule is a prerequisite to enjoyment of life in the country.

Country living offers some dramatic differences in lifestyle relative to the city. In general, rural life is more family oriented. There is less anonymity in the country—most rural residents know quite a lot about their neighbors. Some may find this openness to be a refreshing change, while others might deplore any loss of privacy.

The bottom line in evaluating your own suitability to country life will depend on how well you like the routine. On your acreage you will spend most of your time doing chores and performing seasonal tasks. Much of the time you will be working alone or with another family member. Hence, genuine enjoyment of this life requires that you like the work and that you enjoy your own company.

A final consideration, but one that's usually beyond the control of most individuals, is that of local ordinances and zoning laws. In many areas, the keeping of livestock and poultry is restricted to certain size acreages, and incorporates strict boundary line buffer areas. These requirements, however, may vary widely from area to area.

In some counties at least 2 acres are required before livestock can be kept, and animals must be at least 100 feet from any property line. Restrictions are also placed on the number and kinds of livestock allowed. On the other hand, adjoining counties may have no restrictions at all, and if they do, they may be loosely enforced. Therefore, in order to avoid trouble that can affect plans and hamper efforts, check all aspects concerning local restrictions before engaging in any new enterprise.

Moving from an urban to a rural environment means you want change. Don't overlook the value of close reflection on how much change you're facing. Many people have moved to rural areas, become dissatisfied with their decision, lost time and money, and moved back to an environment similar to the one they left for the rural scene. Hopefully, you will not become one of those statistics.

You are the only one who knows what makes you happy. So before you decide to move and invest your resources, sit back and carefully reflect on the transition to a rural way of life.

PART 2

Acquiring That Spot

Selecting a Community

You know you want to move, but where to? This is the crucial question. By now you should have determined your objectives. There are three steps to finding the right place: choosing the geographic region, selecting a town (or the nearest town), and finding property that you can afford. As you consider these steps, you will ask yourself why you want to move in the first place, how the family wants to live, and what you hope to accomplish on your land.

This chapter will consider these rural ways of life:

- a rural residence;
- a rural residence with expanded growing potential;
- a "mini-farm" that provides you with food;
- a retirement or part-time farm.

The Rural Residence

If your objective is a rural house, your choice of a geographic region will be based primarily on climate (your personal comfort), social and cultural preferences, and job opportunities.

Regions desirable in all these factors attract a lot of new residents. In many areas this means strong competition for the available small-acreage plots. Much of the westward migration in the United States during the twentieth century can be attributed to climatic conditions viewed as favorable by large numbers of people.

As you consider geographic areas, your age and thus your planning horizon is important. If you are nearing retirement, factors such as cost of living and climate will likely outweigh employment opportunities and long-range water and energy availabilities. If you have more time, you will give more weight to where you expect to hold a job, to environmental quality, and to social life and culture.

Selecting your house site depends on such factors as the view, surrounding growth, and the general appearance of the area. Do not overlook availability of a water supply, whether the soil can handle a septic tank, how rainwater drains, and from which direction the wind blows. These are difficult or impossible to change. Unless a common water supply is available, an agreement to purchase land for a homesite should be contingent on availability of an adequate water supply.

The Rural Residence with Expanded Gardening

If you plan to grow much of your food, consider the climate, which restricts what will grow well. Meteorologists define climate as the average weather conditions in a specific place, determined by actual records over a long period. Individual characteristics such as day-length, sunshine and temperature intensity, the frost-free period, and relative humidity combine to determine the quality of the environment for growing food.

Your choice of the garden site is extremely important. Consider the soil depth and texture, the frost hazard, and the direction and degree of slope. Although you can alter the soil composition, you will get a head start on a successful garden by starting off with a deep, well-drained soil that contains substantial organic matter.

Within any general climatic area, frost hazard can vary greatly. The elevation of your garden relative to its surroundings can spell the difference between an ideal site and one that is "frosty." The principle underlying "frost pockets" is that cold air is heavier than warm air, and cold air displaces warm air in low-lying areas. Frost pockets form where cold air cannot drain to still-lower areas. It's critical to understand this if you plan to grow crops that require the entire frost-free period to mature.

Check the slope of your land and the texture of the soil. Few experiences are more disappointing than to find your topsoil (and perhaps the plants that had been growing in it) at the bottom side of your garden after the summer rain you had awaited with so much anticipation.

Aside from the erosion problem, hills are hard to tend, especially if you use power equipment. Try to grow vegetables on a gentle south or southwest slope, where the sun warms the soil in spring and speeds crop growth throughout the summer. Facing south is especially important for plants whose early-season growth is important to their maturity.

For certain fruit trees, however, warming the soil too much leads to early-spring growth that may bring the trees into blossom too soon, while it could still freeze.

The Mini-Farm

On mini-farms, you grow more food for the family, and you give away, barter, or sell the rest. You don't depend on selling produce for your income. On such a farm, you might raise meat, keep chickens for meat and eggs, tend beehives for honey, or milk cows. The climate matters less for livestock than for fruit and vegetables.

Perhaps the factor that most distinguishes the mini-farm from the expanded garden rests in the differing public perception of keeping animals. Zoning rules developed to protect against potential annoyances (sounds and odors) or damage (digging, waste disposal, or chewing) keep animals out of many neighborhoods. Zoning restrictions generally come from town or village commissions, not regional ones.

As you consider the community, take careful notice of emerging changes in settlement patterns. Regardless of the present situation, if more people are moving to the general area—say, to a new subdivision a few miles away—then the area could be changing, and the neighbors probably will be less welcoming of certain kinds of livestock. If your serious gardening neighbors observe your goat pruning their azaleas, or your pig sampling their strawberries, you can be assured they will question whether more restrictive zoning requirements are in order.

If you keep animals, drainage (both surface and underground) will be important. The site must allow for locating your livestock activity so that waste materials cannot be carried to your water supply, or your neighbor's. Prevailing winds should not carry livestock odors to the neighbors.

The Part-Time Farm

The part-time farmer moonlights on the farm while keeping a job or drawing on pensions or investments for the main income. While not necessarily the case, many part-time farms (especially on limited acreages) are expanded versions of the mini-farm. The primary difference may be that production of crops and/or livestock in excess of family needs is intentional.

The extent to which you intend to supplement your primary income with farm earnings will dictate the size and intensity of your operation. That will, in turn, influence both the geographic area and specific site that contributes the most to reaching your objective. Since, as a part-time farmer, you will spend more time operating the farm than the mini-farmer, living near your town job may be an important factor. Time spent commuting takes away from time available for the farm. Consider the seasonal progression of work for the crops or animals you choose, and whether it will mix well with the demands of your other job.

It's important to live and farm near markets where you can sell your produce, near services (financial and informational), and near people who can work for you during the busy times. When you plan to earn money from your farming, you want to spend as little as possible modifying your soil or your site. Look for land with the best natural characteristics for what you plan to do.

Consider Distance

Your choice of community will be based more on personal living desires than on the way you intend to utilize your land. Your background and way of life so far has influenced your expectations as far as public services and facilities are concerned. These attitudes will largely determine your choice of community, so check out the services provided through local taxes. For example, for families with school- or preschool-age children, the public school system may be an overriding factor. If they bus the children, the length and nature of the ride can affect your children's schooling. In many rural areas, the school is an important social institution and you want to be able to get there easily.

Reasonable access to doctors and hospitals matters most to families with small children, as well as to older people. Reaching the doctor is more than a matter of distance. The availability of ambulance service may be of critical importance for certain types of emergencies.

Consider your safety. Most rural areas have high-quality but thinly spread law enforcement services. If you're accustomed to seeing the police drive by often, then the infrequent sight of a county sheriff's car may leave you feeling insecure.

Perhaps of more common concern is fire protection. Before you relocate, find out if it is in a fire control district. This will not only determine the availability of firefighters in an emergency, but also help establish your hazard insurance rates.

Roads can be a major problem in some parts of the country. Mud or snow clogging roads is merely inconvenient if you were going to a party, but spells disaster should you have a medical emergency or a fire. Many rural roads are well maintained, but others are not. Those most likely to receive care are mail routes, school bus routes, and fire access roads. Access to an all-weather road can lessen your feeling of isolation.

Public transportation, if it's available, can be expensive if your trip to work is a long one. If you plan to drive a long way to a job, consider the toll the trip will take on you, and the cost of gasoline, which is very expensive in some areas. If you expect to leave the country for cultural and social activities, as well, don't forget the cost of this. Many people who migrate to those rural settings, however, view their new living situation as its own recreation.

Beyond the availability of services, the cultural, political, and religious environments of an area can influence your choice.

Be Savvy

Several cautions are in order. First, define your objective clearly. Make certain you are moving toward a new well-defined experience, not merely running away from your present situation. Second, be careful of overenthusiasm. The thought of new experiences and challenges is exciting to nearly everyone. Don't let that excitement cloud your vision.

Finally, be cautious of hasty decisions. Advance planning of your land acquisition will be well rewarded when both the location and site characteristics are carefully matched to the objective underlying your relocation decision.

The Internet has made thinking ahead in this regard easier in that you can "visit" and consider the pros and cons of real estate near and far without leaving your desk. Here are some Web addresses of groups that can help with this:

Midwest Organic and Sustainable Education Service. This nonprofit organization promotes sustainable farming in the Midwest. Its Land Link service is a classified ad listing of land available and sought. See http://www.mosesorganic.org/landlinkup.html.

Center for Rural Affairs, Lyons, Nebraska. This organization devotes some of its energy to helping beginning farmers and can help you find land. See http://www.cfra.org/resources/beginning_farmer/land_link.

USDA Farm Service Agency. Occasionally, properties that have been lost due to unpaid farm loans come back into the hands of the USDA. These properties vary from less than an acre to many acres. The buildings are often in poor condition, but such land can be had at a relatively low price. The listings can be searched state by state at http://www.resales.usda.gov/FSA/FSAPropMain.cfm.

Alternative Rural Communities

Communal living in the country did not die at the end of the 1970s. Neither did homesteading, a word people tend to associate with the American West of the late nineteenth century. Both movements still attract people to backcountry tracts in several regions of the United States. Residents of communal farms live and work as loose coalitions. One commune in Virginia houses about a hundred people who grow crops and make hammocks, providing them with a handsome income.

Homesteaders range from those who want to obtain as much food and energy independence as possible to those who want the best of both worlds, and can afford it. The goals and work ethic of the homesteaders do take you back to history. If you seek this life, you might want to provide alternative energy sources and grow organic food.

In years past, rural building inspectors had difficulty enforcing local building codes because many homesteaders wanted freedom to build things their own way. In some areas the people housed in unrecorded, unlicensed structures outnumbered those in regular housing. In California in the 1970s, for example, authorities yielded to the homesteaders and legalized rural buildings with nonflush toilets. In the twenty-first century, few people really want to live that way. Most homesteaders want to provide basic amenities as simply as possible. They are willing to live a simpler life, but they know how to ensure comfort.

The commune population, like a kibbutz in Israel, runs on cooperative work. The residents share decision making, income, gardening, animal raising, and child care. Some settlers who maintain close urban ties through their jobs are not attempting to undertake a subsistence homestead. Depending on their individual situation, they will seek to be within commuting distance and locate near a hard-surfaced highway. They will not be able to personally devote as much time to improving their site as homesteaders or retirees.

Living Arrangements

One of the benefits of communal living can be that a group can buy land as a partnership, pooling their funds. Later, they can split the land among themselves. Check local zoning regulations to be sure this is allowed. Most areas have minimum acreage laws for parcel splits, and limits to the number of residences per parcel.

All people involved in settling on a tract of land should understand their goals. Draw up an owners' agreement or an agreement between occupants to specify the group's goals. They can then select the form of legal arrangement best suited to their situation.

The best strategy for persons seeking to live on a few acres is to locate in areas where there is little competition from commercial interests, land developers, speculators, and recreationists. The price of land will be lowest where there is less active competition for the available tracts, all else being equal. The very act of settling, however, brings gradual growth of demand and increased land values and taxes.

Consider all of the factors any rural migrant would—the land, its fertility, sun and wind patterns, distance to stores and doctors, the growing season, and whether you can have a town job. Of course, if you are planning a commune, you might not consider jobs and money in the same way you used to. On some communes, residents assign hourly wages to work, but the wages are the same for all kinds of work, from tax preparation to sweeping out horse stalls.

The same logic hasn't infiltrated the real estate community, though. Placing a value on your rural land boils down to the market and the assessments. If buyers will pay a lot for a house, farm, or open land, then any other parcels in the area usually find themselves valued higher when it's time for the tax bills to go out.

Legal Considerations

Established communes run as businesses, and new residents contact the owners when there are openings. The experience can be like moving to a camp, permanently. These organizations have their own legal setups, and you must learn what they are, and understand what you're signing.

When you are part of a group starting a new communal farm, there are three basic ways for a group to purchase and occupy a tract of land.

These are tenancy in common, joint tenancy, and incorporation. Each form of joint ownership has advantages and disadvantages.

With tenancy in common, each purchaser shares an individual interest in the total parcel of land, and it may be in an unequal interest relative to the others. Each party's land is specified in a contract among all the owners. If any co-tenant dies, his interest in the property goes directly to his heirs.

In joint tenancy, each person has an equal interest in the land. If an owner dies, his interest in the property goes to the surviving owners, not the heirs. This form is not permitted in all states.

A corporation has all the legal aspects of a single person. The advantage is that shareholders do not have personal liability for acts of the corporation. The main disadvantage is the high cost of incorporation.

Making Your
Final Choice

You are ready to select and purchase rural property. By now you have learned that you should take the family to look at the land, know what you want to do with it, and consider the greater community. But there are other pitfalls to avoid. Ask yourself:

- How often can you realistically expect to visit the property?
- Will acquisition of this property result in a major change in recreational patterns of your family?
- Will job-holding members of the family still continue to work in the city?
- Will it cost a lot to commute to work?
- If energy costs go up, how will that hurt your budget?
- What will be the effect of the move on the time jobholders spend with the family? (Many exurbanites who move to the country hoping to spend more time together and strengthen family ties have experienced the opposite result.)
- Do you have enough cars or time to chauffeur children to activities?
- What is the effective rate of interest you will be paying on any mortgage on the property?

Land Factors

One of the first items to check is the parcel description. Later, at the closing, you will probably want to obtain the services of a title attorney or other professional title examiner. Initially, however, examine the recorded plat, certified survey map, or last deed recorded on the parcel. Compare the legal description with the property itself. Familiarity with the legal description of the parcel will be helpful as you compile information from various government offices about specific parcels you are interested in.

The county recorder's office usually can provide you with the legal description of land parcels. Commercially prepared plat books also contain property descriptions and can be found in the local library, assessor's office, or the county Extension office.

Your next stop should be the local zoning office. This will usually be a county or township office. Discuss your plans for the land with the local zoning administrator. If the land isn't zoned for what you want to do on it, you will have to file paperwork to change the zoning, or obtain an exception.

Ask yourself if zoning on adjacent properties will conflict with your interests. If the land is in an agricultural zoning district, is the zoning used of the cumulative or exclusive type? Since almost any land use is permitted in cumulative agricultural zoning districts, this type of zoning is relatively ineffective in dealing with many land-use problems.

What are the applicable zoning regulations affecting building location on the parcel? Especially note required setbacks from the street or road, side yard and rear yard lines, as well as from septic field locations. This is particularly important if you have a specific house plan in mind. Early efforts to consider fitting a proposed house to the parcel, rather than fitting the parcel to the house, will result in a house and lot that complement each other, as well as being compatible with garden, patio, driveway, and recreation areas.

Are there special zoning regulations that may affect use of the parcel? (Think about restrictions on mobile homes, livestock, and residential development adjacent to navigable streams and lakes, and areas that are subject to flooding.)

Many areas of the country also are covered by separate ordinances specifying requirements governing the creation, transfer, and permissibility to build on a parcel. Information on these matters can be obtained from the zoning administrator, local or regional planning commission, and the recorder's office.

Septic Systems

Waste disposal is one of the most basic, and expensive, needs of country living. Municipal departments will advise of the regulations, usually set by the state or county. If there is a house on your new property, ask about the septic system's age, likely condition, and size. Adequate size is important if useful life of a septic system is to be maximized.

Septic systems are "sized" based on number of bedrooms and type of plumbing fixtures in the home. Beware if you are taking your entire family into a house where one older person lived for the past few decades. You probably will have to upgrade or replace the septic tank.

You might want to obtain information on soil type and soil permeability, slope, depth to groundwater, and depth to bedrock—all useful facts in determining how well an existing or new septic system will operate. The officials will tell you whether your land can support a septic system on its own, or if you have to alter it.

Before you make an offer to purchase, determine that the parcel has passed any required percolation (called "perc") tests. If this has not been done, any offer to purchase should be contingent on the parcel passing such tests.

Actual installation often requires additional permits and inspections by local officials. Learn your town or county's regulations for periodic pumping of sludge from the septic tank. This extends the life of your system but can cost about $100 or more.

Most rural areas obtain their water supply from a well. Sometimes you will share a well with other homes. Wells can cost thousands of dollars to install. Be prepared for periodic maintenance costs on pumps and related equipment. Also, periodic testing of group wells is often required by health officials, and this testing cost is often charged to the user.

Land-Use Changes

You might consider your new home an idyllic country spot, but in reality, development may be galloping across the landscape. Are you prepared for the changes you might witness after you relocate? Many people who live in semi-rural areas expect development to stop after they arrive. This doesn't happen.

Your final review step, then, should be to determine what kind of community you are choosing, what changes have occurred there during the last few years, and what changes will likely occur in the next several years. Talk to people in town and at the municipal offices. Drive around and visit with potential neighbors and other local residents.

Ask zoning officials what kinds of rezoning changes are being made, where the changes are typically occurring, and the frequency of such changes. Look for postings about new subdivision hearings. Remember that if you have little difficulty in acquiring your land, others are doing what you're doing. Therefore, you may soon have several new neighbors.

Evaluate carefully the impact of the political subdivision location of the property. Often there are significant differences in level of services provided, as well as tax rates, between neighboring townships or municipalities. Keep in mind that many services such as garbage collection and snowplowing, which are included as tax-paid services in urban areas, become the responsibility of each landowner in rural areas.

Getting the Parcel

Your major concern should be to avoid any surprises, during or as a result of the purchase and closing process. With proper planning and preparation, the two most important kinds of surprises—those relating to quality of the title and costs of transfer—can be avoided. In many cases it is best to obtain the services of an attorney to assist in preparing the offer. Once an offer is signed by both buyer and seller, it becomes a binding contract on which the entire transfer process rests.

In relatively simple transfers, or if the buyer is very familiar with the transfer process, he can prepare the offer himself. Usually, if a broker is handling the sale, he will be willing to assist or to prepare the offer. If you intend to hire an attorney for the closing, the additional charge to prepare the offer is usually well worth the cost.

The offer must be in writing to conform to the Statute of Frauds. Amount of detail in the offer will depend on the transaction's complexity. The purpose of the offer is to specify what is being bought (and sold), what the price and other terms of sale are, and who is responsible for the various costs that result from the transfer process itself.

You can specify in the offer what costs you believe the seller should pay. Whether he will agree will depend on a variety of factors, including how long the property has been on the market, the price, and the quality of the property.

PART 3

Improvements for Your Place

Remodeling a House— Will It Be Worthwhile?

A house on small acreage may offer comfortable living conditions at a moderate cost. To meet your needs and desires, it probably will require remodeling. Perhaps it is old and run-down, too small, or otherwise not quite what your family wants.

Keeping the older house has real advantages. It may save money, because rebuilding is expensive. Remodeling is also a conservation measure, for it doesn't use all-new building materials. If the house has a unique and desirable character, saving this example of our architectural heritage is another plus.

But every house isn't worth remodeling. How do you evaluate the house, and how do you assess your needs and plan to satisfy them? Give it a thorough inspection. Make your own observations, and get some professional help. Consider these problems:

The Foundation

The foundation supports the entire structure. Any foundation failure may distort the house frame, resulting in problems with doors and windows, loosening of siding and interior finish, and cracks that allow air to blow through the house.

Most foundation walls of poured concrete have hairline cracks that have little effect on the structure. However, large open cracks indicate a failure that may get progressively worse, so some professional guidance may be in order. Crumbling mortar in brick or stone foundations can be repaired, but if most of the mortar has deteriorated, a major repair is required. Complete replacement of the foundation may also be necessary.

Localized failure or minor settling may be corrected by releveling beams or floor joists. If pillars are used under the house, or under porches, check to see that they are sound. Sometimes replacement is more difficult than it appears.

Standard Wood Frame

Examine the building frame for distortion from failure of the foundation or from inadequate framing. The general condition may not be too obvious, but look for decay or insect damage.

Look at the floor supports from the basement. It will be difficult if the basement is a crawl space. In a basement, wood posts should be examined for decay at the juncture with the floor. Girders resting on these posts should be checked for sag by sighting along the girder. Some sag is quite common under heavy loads such as bathtubs, heavy appliances, or partitions. This sag usually affects appearance more than strength, but it can be critical if the floor above slopes noticeably.

Sill plates or joists and headers rest on top of the foundation and thus are exposed to moisture from the concrete. Wherever possible, examine these contact points for decay and insect damage. If the basement or crawl space seems very damp, the entire floor framing system should be examined.

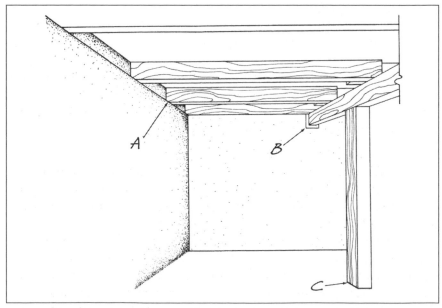

Check wood for decay at points of contact with concrete, such as: A, floor joists supported on concrete walls; B, framing supported in a pocket in a concrete wall; and C, wood post supported on a concrete floor.

Sag in floor joists is not critical unless it is readily apparent. A more common problem is to note springiness as you walk across the floor. The floor can be firmed up by adding extra joists or girders to increase stiffness. A point of particular concern in the floor system is framing around stair openings. Check floors around the opening for levelness. Where floors are sagging, the framing should be leveled and reinforced.

Wall framing usually has more-than-adequate strength, but may become distorted if the foundation settles too much or floor framing is inadequate. Check doors and windows for squareness to make sure they do not warp. Also check for sag in headers over wide window openings or wide openings between rooms. Headers that sag noticeably will have to be replaced.

Look at the roof for sag at the ridge, in the rafters, or sheathing between rafters. If the ridge line is not straight or the roof appears wavy, some repair may be necessary.

Siding, Windows, and Roof

Exterior wood on a house will last for many years with reasonable care. Excessive moisture can ruin paint, but so can poor surface preparation, poor paint, improper application, or incompatible successive coatings. Regardless of the cause, you may have to completely remove the paint before repainting.

When examining the siding, look for space between horizontal siding boards by sighting along the wall. Where warped boards leave big gaps, new siding may be required; however, if boards are not badly warped, renailing may solve the problem. Check the ends of siding boards for decay where two boards butt together, at corners, and around window and door frames.

Good shingle siding appears as a perfect mosaic; worn shingles appear ragged, and individual shingles often are broken, warped, and upturned. New siding will be required if these shingles are badly weathered or worn.

You can grout cracks in brick or stone veneer, and repoint the joints, but large or numerous cracks may be unsightly even after repairs. To prevent water from entering masonry walls, examine the flashing at all projecting trim, copings, sills, and intersections with the roof. Plan to repair any of these places where flashing is not provided or where it needs repair.

Check all windows for tightness of fit and examine the sash and sill for decay. Weatherstripping can better seal the window, but if the sash or sill is decayed, that part must be replaced. If you plan a replacement, however, measure the window and determine if it is a standard size. If

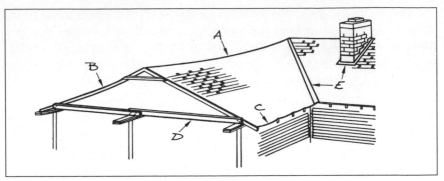

Watch for sag at *A*, ridge; *B*, rafters; or *C*, sheathing. Rafters are frequently tied, as at ceiling joist, *D*, to prevent them from spreading outward. Flashing, *E*, is used at intersections of two roofs or between roof and vertical planes.

not, the opening will have to be reframed or a custom sash will have to be made—both of which are expensive.

In cold climates, all windows should be double glazed or have storm windows. Again, if the windows are not a standard size, making up storm windows may be expensive.

Severe roof leaks should be obvious from damage inside the house. But general condition of a roof that is apparently not leaking will be more difficult to determine. If possible, look carefully in the attic for any signs of problems.

On the roof, scrutinize the condition of the shingles. Asphalt shingles show deterioration by becoming brittle and losing surface granules. More important is wear in the narrow grooves between the two shingles, which may extend completely through to the roof boards. Where such signs of wear exist, new roofing is required. Wood shingles need to be replaced when numerous individual shingles are broken, warped, or upturned.

Built-up roofing on flat or low-sloped roofs shows wear by bare spots in the surfacing and separations and breaks in the felt. Bubbles, blisters, or soft spots also indicate the need for major repairs. Examine the condition of the flashing; corroded flashing should be replaced.

The Inside

Interior surfaces may have deteriorated due to wear, distortion of the structure, or the presence of moisture. Wood floors should be checked for buckling or cupping of boards. If cupping or buckling is not too severe and boards have not separated excessively, the floor may be returned to good condition by refinishing. But first, be sure the floor is thick enough to withstand sanding.

Floors with resilient tile should be examined for loose tile, broken corners, cracks between tile, and chipped edges. If any tile must be replaced, the entire floor will probably have to be redone because new tile will seldom match.

Plaster on walls and ceilings almost always has some minor cracks, which can be patched. If large cracks or holes are numerous, a new wall or ceiling covering may be required. Bulging or loose plaster also indicates the need for new walls or ceiling.

If walls have more than two or three layers of wallpaper, remove the wallpaper before painting or putting on new wallpaper. Paint also may have built up to excessive thickness on walls and ceilings, or be badly chipped. In either case, you should completely remove old paint or apply another wall surface.

Trim can be refinished. However, if the old trim is badly chipped or cracked, remove it. Ornately carved designs are the most difficult to refinish even though they may provide the most rewarding results. If replacement of some trim is required, new sections can be custom-made. However, this may be costly.

Decay and Insect Damage

Look for decay in any part of the house where wood has remained wet for a time, such as close to the ground. Decayed wood can be identified by its loss of sheen, abnormal color, and sometimes fungal growth on the surface. A test for extent of decay is to prod the wood with a sharp tool to see if it mars easily. Pry out a splinter. If toughness has been reduced, the wood may break across the grain with little splintering and lift out with little resistance. Sound wood will lift out as one or two relatively long slivers, and breaks are splintery.

Termite damage is sometimes harder to spot. Look on foundation walls for earthen tubes. The tubes are evidence of subterranean termites, which use them as runways from soil to the wood above. These termites follow the grain of the wood as they eat their way through, leaving galleries surrounded by an outer shell of sound wood.

By contrast, nonsubterranean termites, which are found only in warm coastal areas, eat freely across the wood grain. The nonsubterranean type requires no connection to the ground. If any termites are in evidence, get the opinion of a professional exterminator.

Insulation

Check the ceiling for insulation. If there is little or none, your first priority would be to add ceiling insulation. Ask your local utility company for guidance on how much to add, as well as for general guidelines on energy conservation. But as insulation is added, good attic ventilation becomes critical. Floor insulation is also a good investment.

Storm windows help by reducing heat loss through glassed areas by up to one-half, and also reduce air leakage around the windows. Weatherstripping doors and windows, as well as caulking all cracks, costs little for materials but does require considerable labor.

The next step would be the walls. Where insulation is added to the walls, greater expense and effort is involved because this must be professionally applied. The finished product may be worth it, however, in the resulting comfort. In areas of cold winters, some precautions are necessary to prevent moisture problems in the walls.

Moisture Control

When moisture-laden air can leak through cracks to the outside of an old house, moisture is seldom a problem. But as the house is tightened by weatherstripping and storm windows and doors, the air is no longer carrying off that moisture. A vapor barrier, placed on the warm side of the wall, prevents that moisture from simply moving into the walls and condensing there as it meets the cold outside wall. Lack of a vapor barrier commonly results in exterior paint peeling problems.

Several coats of oil-base paint on plaster provide some resistance to water vapor. If relative humidity in the house is kept low, this paint may be an adequate vapor barrier. Additional vapor resistance can be added by applying a highly moisture-resistant paint or vinyl-coated wallpaper. Where new panel material is planned as interior finish, a good vapor barrier such as polyethylene film can be placed over the plaster first.

Crawl space moisture often migrates up through walls and living space. Moisture levels under the house can be reduced by good ventilation, but a vapor barrier ground cover is even more effective because it prevents ground moisture from coming up into the crawl space.

Utilities

In this time of concern for saving fuel, the efficiency of the heating system is equally as important as the insulation for the house. Utility companies can advise you on this. If the heating plant is quite old, a new one almost certainly will pay for itself in improved fuel economy. Look at the general condition of the furnace or boiler, but seek professional help for a thorough inspection.

Where firewood is available on your acreage, consider using it for at least a secondary source of heat. Wood-burning units that can be attached to the furnace are available, and wood-burning stoves are quite efficient. The problem is that you have to keep feeding them. Even if you try to use wood as a primary source, consider having an oil unit as a backup.

Plumbing and Electricity

Check several faucets to see if the flow is adequate and if there is good pressure. For a private water source, the gauge on the pressure tank should read a minimum of 20 pounds, preferably 40 to 50 pounds. Have the water in any private well tested.

Look for leaks in the water system. Rust or white or greenish crusting on pipes or joints may indicate a leak. Check for clogged drain lines by flushing the toilet and observing any sluggishness. Also run water for a few minutes to determine if the drain lines are clogged. Where a new drainage field is needed, some codes require percolation tests of the soil.

Electrical service should be at least 100 amperes for a moderate three-bedroom house, and probably 200 amperes for a larger house, or if you will use air-conditioning. New service with a larger capacity will be a major expense.

Some wiring is usually exposed in the attic or basement where it can be checked. If cable insulation is deteriorated, damaged, brittle, or crumbly, or if armored cable or conduit is badly rusted, the wiring should be replaced. Also notice if there is at least one electrical outlet on each wall of a room and if ceiling lights have wall switches. If these are lacking, consider the expense of bringing the wiring up to current standards.

General Considerations

Consider the arrangement of the house and the changes that may be required for convenience. Look at the rooms in terms of furniture placement and adequate circulation space. Also consider storage requirements. Many older homes have few, if any, closets. Do not be bound to traditional uses of rooms, but look at spaces in terms of your way of life. Sometimes moving a door or eliminating a partition may do wonders. However, keep in mind that partitions may be supporting the floor or roof above.

Appearance of a house is largely a personal matter, but some general guidelines should help. The well-designed older home often has a desirable character. Keep changes to a minimum.

Unity of design is generally a governing factor. Windows and trim should match each other and the house's style. Porches and garages should blend with the house rather than appear as attachments. Too many types of siding, or ornamentation that appears stuck on as an afterthought, may present a confused appearance. Also consider what can be done with paint and landscaping. No house looks good in a rundown condition or with unkempt grounds.

Summing Up

Foundation and framing are the most critical items. If the foundation is good and framing is generally square, the house is probably worth remodeling. However, if numerous other repairs are required, or the house generally does not meet your needs, remodeling may be too difficult. In the final analysis, make a detailed cost study or get a contractor's bid to help you decide if the remodeled house is worth the cost.

Assess Family Needs

Your present family needs go beyond the number of bedrooms in the house. Consider your living habits, work, relaxation, and entertainment; and think of the needs of each in terms of space, privacy, storage, access to other areas of the house, and comfort level required for each activity.

Entertaining large groups requires large rooms. If you frequently have overnight guests, a spare bedroom may be important. If you plan to raise your own food, the kitchen and utility area should provide space for the processing involved, and storage of canned or frozen food.

All families change. A young, growing family undoubtedly will need more space in the future. Look at expansion possibilities. Consider using presently unfinished space such as attic, basement, or attached garage. Generally there should be room for an addition to a rural house. The main thing is to plan such an expansion or addition at the time you remodel. Size the heating system and electrical service for add-ons. Rough plumbing may also be planned for expansion.

Of course, a large family decreases in size as children establish their own homes. Perhaps you might see if an area of the house could be closed off except when guests are present. Or, could a part of the house be converted to an apartment suitable for rental? By making the division between units with removable panels, the house could be changed. Of course, the feasibility of this arrangement depends on being in a location where there is a demand for such rentals.

Getting Ideas and Help

In these days of internet databases, it might be harder to winnow out the useful than to find free advice on how to renovate and build. Search for links through your state's Extension Web site. The local library should have books and periodicals on ideas for remodeling as well as how-to booklets. Check also with the lumberyard or building supply dealer and Extension office. To accomplish the actual construction, you have the option of hiring a contractor, doing the work yourself, or a combination of these two. Much depends on your abilities, available time, and extent of work required.

Hiring a contractor does offer some advantages. The work can be completed more quickly by an experienced crew working full time. Financing may be more easily arranged under these conditions, and you could be enjoying your completed house at a much earlier date.

On the other hand, it may be quite feasible to do much of the construction yourself. You will have to evaluate your own abilities and make the final determination. Projects go very slowly when worked on by one individual in his spare time. If the house is to be occupied immediately or at the earliest possible moment, do the necessary items at once and plan to work on other projects later. Nobody wants to live in a continual mess, and nobody can work continuously without the risk of having the project go sour.

Be as realistic as possible. It will increase the enjoyment of doing the work, as well as your ultimate satisfaction with the finished home. And, finally, realize that your ideas may change.

A Personal Experience

By Richard A. Biggs

Having grown up on a family farm, I knew that when I chose a place of my own it would have to be in the country. When you're used to the sight of a fox darting across a snowy, moonlit landscape, or the sweet aroma of new hay being packed away in the barn, or the sound of frogs croaking at the arrival of spring, it is simply impossible to consider any other way of life.

Soon after we were married, Nancy and I had the opportunity to purchase an old house and a few acres. The house had not been cared for properly for about fifteen years, and needless to say, was in rather poor condition. There was no heating or septic system, the water pipes had all burst during previous winters, only three electrical outlets worked, windows had been broken, and the front porch floor had rotted into oblivion. The lawn had become a tall pasture and the shrub border consisted of lilacs supported primarily by honeysuckle, brambles, and poison ivy.

In thinking back, I believe the main reason we fell in love with this place was that it had been neglected so long. We realized that every single project we undertook would be a fantastic improvement. By squinting our eyes very hard, we could picture what the house and lawn must have looked like at one time, and we became determined to bring it back to life again.

Three years went by while we both worked to save money for the much-needed remodeling. On weekends we would drive up to our "new" home and work out in the yard. At one point, I recall thinking that maybe we ought to slow down a bit, as there wouldn't be much left to do when we actually moved in. What a joke! After six years living there, we decided we'd never really finish restoring the place.

At the end of our three-year wait-and-save period, we decided it was foolish to delay remodeling any longer. The cost of building materials was rising much faster than our savings balance! We contacted a local builder who gave us much-needed advice on soundness of the structure, major jobs that must be done, and approximate cost. Nancy drew up a rough plan of the changes and additions we desired, and my father, being an architect, drew up the final plans.

Saturday Meetings

To help keep remodeling costs down, our builder agreed to meet us each Saturday morning and advise us on how and what to start removing. This consisted of such jobs as tearing out walls, small closets, and the four brick chimney flues. These various tasks took several months, and finally we were ready for the carpenters, electrician, plumber, roof man, and heating system installers.

Before long we had a five-zone heating system, new electrical wiring, a septic system where the garden used to be, insulation and storm windows installed, a new front porch, and new wallboard tacked over the existing horsehair-plaster mix. Unfortunately, things did not go exactly as we had anticipated, and our remodeling costs began to soar way over the initial estimate.

When the roof man came to repair broken and missing portions of our slate roof, he found that each slate had weathered so much that it was impossible to replace one without splintering each surrounding one. Naturally, the only solution was to install an entire new roof.

On another occasion, we accidentally discovered a large colony of termites when I fell through the kitchen floor. This led to a rather extensive wood replacement bill.

The major remodeling took four months before we were finally ready to move in. There were still several inconveniences to contend with, such as the 8-foot hole in the wall where the fireplace would eventually go. Fortunately it was springtime, and these conditions were corrected by midsummer.

Our white frame house sat on top of the windiest hill in the area, and heating costs are phenomenal. We thought we had taken enough heat conservation measures, but it turns out we were sadly mistaken. Additional insulation, plastic over the storm windows, and one hundred pine seedlings as a future windbreak all help, but we still have a long way to go.

Termites were a constant problem, and after unsuccessful trenching and treating of the house myself, we finally had a professional exterminator in.

The Four-Year Paint Job

The idea of living in a large frame house with various shingle designs and intricate scrollwork appealed to me—until I started painting the outside. It took four summers to scrape, sand, and put three coats of paint on the entire house. When I finally finished the last side, the first side had already begun to peel.

Our driveway was about $1/4$ mile long and consisted of dirt, rocks, and gullies. It got to the point that when it rained, friends called first to check on our driveway condition before they come to visit.

When you live in an old house surrounded by fields of wheat and corn, you have to be prepared for occasional unwanted visitors. During our first winter, a rather large rat decided to spend the cold months with us. After several unsuccessful nights of trap setting, and Nancy's firm decision that it was either the rat or her, I set out poison bait. According to the directions, the rat was supposed to go outside to expire. Unfortunately he didn't read the directions, and it took several weeks and many cans of air freshener before our living room was enjoyable again.

While we had several disheartening experiences in our "new" home, the happy times far outnumbered the unpleasant ones. Each year we added a few more plants to the landscape, finished off another room in the house, added a bit of gravel to the driveway, and harvested a little more fruit from our young orchard.

Occasionally, we even had time to stand back and admire our home and realize that we had something very special indeed.

Building That Dream House: Don't Be Caught Napping

Building a new home can be one of the most rewarding experiences of living on an acre. What could be more enjoyable than seeing the dwelling that you have carefully planned and designed take shape and become the home that you and your family will enjoy and be proud of for years to come? Without careful planning and work on your part, however, your experience can deteriorate into a trauma, resulting in a house that is less than your dream: poorly planned, too expensive to build and maintain, or one for which you have signed binding legal instruments not in your best interests. Your entire investment could be jeopardized by neglecting to adequately protect yourself through careful selection of site, builder, and lender.

The first step in planning a home is to determine the size and design needed for your family. Unless you have unlimited funds or a rich uncle, space not used is an unnecessary initial expense and will require time and money to clean, maintain, and heat over the years. The size of your family is probably the primary factor used to determine how large a home you need.

In addition to size, you should also be concerned with the age of the family. A young growing family will need room to expand, while older children will soon be leaving the home. Special family interests or hobbies should also be considered when planning the size and design of your home. Give careful thought to the needs of older or handicapped family members. Just a little effort in this area, such as wider interior doors, single-level designs, special features in kitchens and bathrooms, and ramps, when necessary, can certainly enrich their lives. Proper room arrangement will enhance the livability of the home by allowing for adequate traffic flow and separation of activity areas from sleeping space.

Because you want to live in the country, you should want to build your country house as sustainably as possible. It would go against the spirit of homesteading to order a log house kit made from illegally harvested lumber from Asia, or logs that traveled thousands of miles to reach your building site. These are the realities of twenty-first-century house building.

You may prefer to consult a professional architect, who will be able to provide valuable technical expertise in design as well as structural features. The architect can also help with selecting a contractor and preparing the contract. The cost of architectural services will vary according to the dwelling's size and cost and the extent of the services you request.

Think Sustainably

Your goal should be to build using materials obtained locally, if possible, to avoid the energy expenditure of transporting wood, stone, or other materials thousands of miles. Building green isn't just a nice idea in the twenty-first century; for someone wanting to move to the country, it should be a priority. Energy is too expensive now.

Once you start to ask how to build sustainably, house designs that at one time you may have associated with hardscrabble pioneers or native peoples in Third World countries today emerge as modern, even state-of-the-art building. Some of these materials and techniques are so odd looking (by the standards of the North American middle class, perhaps) or require so much care that we don't mention them here. But the viable techniques you should seriously consider include the following:

- Straw bale houses can use local straw and provide efficient, long-lasting structures. Straw holds few of the seeds of the original grasses and so lasts a very long time without attracting animals or molding.
- Papercrete block houses use building blocks made of waste paper mixed with cement. Inventers in New Mexico and Colorado have had success with these structures.
- Earthen houses, called cob houses, or adobe, have walls made of a mixture of soil, sand, and straw. You would need to devote the time to massage this material into the sculpted walls it forms or to find a contractor who specializes in this sort of building. Experts say the material is similar to stone when it dries. It insulates well because the walls are thick. Such houses can be found around the world in a variety of styles, from English cottage to simple rounded hut.

Standard Plans

Standard building plans may be used as an alternative to plans specific to your site. Many firms produce and sell such plans at a moderate cost. Booklets containing floor plans and sketches of many models may be purchased or obtained at building supply dealers or in the magazine section of your newsstand. Standardized building plans are normally available in practically any size or design you might prefer.

Factory-built houses are another alternative you may wish to consider. A factory-built unit is simply a dwelling that is built in sections in a factory, transported to the construction site, and erected on the foundation. The components, which are built under close supervision in the factory, range in size from 4-foot wall sections up to one-half of a modest-sized ranch house.

Companies that manufacture these units usually have several models a purchaser may choose from. These models normally reflect different room arrangements, size, exterior and interior finish, as well as different styles of architecture.

Another important factor to consider in designing a home is the amount of energy required to heat and cool it. With today's skyrocketing fuel costs, the added initial expense of constructing an energy-efficient dwelling will be returned several times over in utility savings. For suggestions on how to make your dwelling energy-efficient, contact your local utility suppliers.

Room for Expansion

Special-purpose features should be carefully considered in planning the home. A carefully designed, well-laid-out kitchen can literally save miles of walking over a period of a few years. In addition, a bright and cheerful kitchen will make some of the routine kitchen work much more pleasant. The addition of kitchen built-ins, such as dishwashers, garbage disposals and compactors, and ranges are a special treat if they do not stretch the budget too far.

Young, growing families on limited budgets may be interested in unfinished basements or second stories as a means of expanding to meet their future needs. Design of the dwelling may also allow for adding bedrooms, baths, or a family room without upsetting the basic room arrangement or causing unnecessary expense.

The Right Site

Selecting the right site is just as important as selecting the style or size of the house. If you already own land, then you must choose the best site available, provided your land includes one or more acceptable sites. If you do not own land, or if your acreage does not offer an acceptable homesite, there are some important factors to consider before buying land.

Remember that local zoning and building codes affect where you can build. Consider the availability of water and septic systems as well as health regulations.

The soil of the site is extremely important. Heavy clay soils are subject to swelling when wet, and shrinking when dry. This swelling and shrinking can break foundation footings, cave in basement walls, and buckle walks and driveways. Unstable soils may slip, if on a slope, causing a building to literally pull apart. Check with government officials who understand these problems.

Many people have made the mistake of not matching the site and design of the dwelling. Ranch-style houses on hilly terrain result in excessive land-leveling costs. Split-level homes built on flat sites require artificial hills, which are expensive and look out of place.

Excessively high water tables can cause a multitude of problems. Occasionally, high water tables are not noticeable except during a rainy season; so check with the adjoining property owners if you are looking at sites during a dry period. High water tables may eliminate the possibility of a basement. An inoperable septic system, probably the most frequent problem encountered when building in rural areas, is often the result of a high water table or heavy clay soil.

Remember privacy and traffic noise, two important factors. You want privacy, but, on the other hand, living in a dwelling located far off a public road may have several disadvantages. Driveways, even if not paved, are expensive to build and maintain. Snow removal may be a problem and an added expense you haven't counted on. Many utilities such as water, sewer, gas, and electricity are more costly if the dwelling is located too far from existing services. Properties located on paved roads with existing water and sewer lines may become heavily trafficked and lack some aesthetic value.

Final Plans

Once you have selected your site, you can turn your attention to developing final construction plans, selecting materials, and developing construction specifications. Consider the general steps, because how you make final plans can vary.

If you choose a factory-built house, your task is to select the model and choose the options. While you have fewer choices, you do have the advantage that construction can probably be completed sooner than if you choose a custom-built house.

If you are planning to use standard plans do some further checking before spending your money on a set of plans that may be impractical. Unless you are intimately familiar with construction details and building costs, you may wish to take the floor plan and sketch to a builder or building supply firm for their comments and approximate cost estimates.

Although a builder cannot bid without complete plans and specifications, he can roughly estimate the cost. With such information, you can look for a plan that closely fits your budget. If you expect to borrow money, you may also wish to show the floor plan to your prospective lender.

When you have decided on a building plan, order at least three complete sets, one for the contractor, lender, and you. Record any revisions on all three copies.

An architect's plans allow you the most freedom. A good architect should be able to give advice about cost and durability, but an architect must understand what you really want. Make sure you have thought things through.

Materials and Specifications

Selecting the building materials and completing the specifications can be an exciting experience. However, it takes time and exploration. Every basic building material of the recent past now comes in many alternatives. Footings, siding, decking, roofing, even framing, come in a variety of natural and recycled materials.

As architect Duo Dickinson of Madison, Connecticut, explains it: "Truly green siding is local, indigenous material, usually of compromised quality to the industrially grown and harvested stuff from the Pacific Northwest (cedar) or made in a factory in the South (fibre cement)." This siding, which is considered "green" because it comes from sustainable small caliper new growth trees transported from within 50 miles of the building site, is typically pine and will require some kind of preservative coating.

Dickinson said that cement board siding is renewable and uses a reasonably low amount of energy to make but expands its carbon footprint because it typically travels such long distances. "Controversially enough, I think vinyl may be truly green," not because it is made of vinyl but because it never needs coating and therefore leaches nothing into the ground as any coated surface does.

Consider alternative roofing products, their cost, and how long they last. For instance, aluminum requires a great deal of power to create, and so Dickinson does not recommend it in most cases, although it does last a very long time. Architectural asphalt shingles waste 30 percent of their weight and material for decoration, so he recommends three-tab asphalt shingles.

Trim materials are undergoing a revolutionary transition, Dickinson says, to PVC. It lasts far longer than wood trim which, today, often is made of low-quality, knotty cedar or rainforest wood brought many miles. It makes sense to salvage redwood or cedar from torn-down buildings, he says.

Begin your search for materials at your local building supply dealer, where you will find samples of various materials. The dealer can describe completed houses made of these materials.

Be cautious about using material you have to order specially, because it could delay the start of construction. The building supply dealer can furnish you with cost information, which you should consider before completing the specifications.

Do not attempt to complete the specification sheet without technical assistance unless you understand construction terminology. The specs will become part of the construction contract, which is a legally binding instrument between you and the contractor. Since the contractor bases his bid on these specifications, it will cost more money to correct any omissions.

Complete each item in the specification form. Give careful attention to special items such as kitchen appliances and fixtures, bathroom fixtures, floors, and wall coverings. Check which sources of heating fuel are available. Depending on local prices and availability, some fuels may be significantly cheaper to use than others. Solicit the advice of local builders, heating contractors, and even the utility companies before making a final determination.

A "lock and key" contract is by far the most popular means of construction and for most of us, the best way. This is a total contract in which the contractor completes all work from site preparation and footings to completion, generally including basic landscaping or at least the finished grading. Upon completion and final acceptance and payment, the contractor will present you with the keys, and you can move in.

Be careful when selecting the contractor. If you plan to solicit bids from several contractors in your area, ask them for references and inspect homes they have built. You may want to talk with previous customers and consider their recommendations. Find out if the contractor guarantees his work and materials, and if he does, see if he honors this warranty.

An alternative to a "lock and key" contract is to build the house yourself, with the help of carpentry labor and subcontracts for some work, such as electrical, plumbing, heating, and cooling. Before you seriously consider this route, be aware of several absolutely essential factors.

First, you must have sufficient expertise in the construction business to at least supervise the construction. You should have enough time available to spend several hours each day at the construction site, and you must have the time to select and buy material. Remember, also, that contractors are regular customers of materials dealers and subcontractors. They can usually buy cheaper than you or I can.

Work and Storage Areas Outside

In the country, you will need more outside storage space for equipment and products for work and play. You must store cars, perhaps a pickup truck, yard and garden equipment (such as a tractor, mower, gasoline, rakes, shovels, fertilizers, and the like), home maintenance equipment (such as paint, ladders, tools, and window screens), spare tires, power saws, lumber, supplies for your livestock, and more.

List what you must store, especially vehicles. Keep in mind that you might replace things as they become obsolete and your needs change. Plan for this, as well as for your changing interests.

Study available plans and visit other families who have already built their homes and are busily engaged in living on small acreages. Storage space for keeping products when they are not being used should be convenient, safe, and adequate in size and shape.

Judging Convenience

The stored item should be easy to see, reach, grasp, remove, and replace. Don't stack articles unless they are of similar nature, such as firewood or bags of mulch.

You can do half an hour's worth of gardening if the necessary machines or tools are easy to reach. If they aren't, you might put off the job.

Bicycles, wagons, and other toys should generally be easy to reach. Storing articles near where they are first used also adds to convenience. Machines removed from storage areas often require routine maintenance before being taken outdoors for use. This implies storing the maintenance products so they can be used on the machine between the storage area and outdoor work area.

Some machines and tools require space for maintenance or repair. In these cases there should be a space that can be cleared so they may be worked on, preferably in the workshop area. If the workshop is in the garage, the car may have to be put outside to leave floor space for repairing a yard machine.

Tools associated with the workbench should probably be stored in a locked cabinet above the bench. Seasonal equipment, for warm- or cold-weather use, can be stored in a less convenient area in the off-season and in a more accessible area during the season of use.

Items can sometimes be interchanged seasonally in the same storage spaces. An example: lawn-mowing and snowblowing machines. Skis and croquet sets are seasonally used and could justify more security in their storage.

Utility connections are needed to operate or service some machines, tools, or products. Locate these conveniently in relation to the area where they are used. Adequate lighting at work and service areas is recommended. Utilitarian lighting fixtures need not be expensive.

Storage should not be too deep. Most items requiring storage do not require more than 12 inches in depth.

Safety Aspects

You should be able to remove and replace machines and articles without hurting yourself or damaging the objects. Lighting should be good. You need adequate ventilation in work areas where carbon monoxide, dust, or noxious fumes may be generated. An exhaust fan can be mounted in the wall or ceiling.

Children are unaware of hazards that exist in such work and storage areas. Responsible adults can keep hazardous items in locked or inaccessible storage spaces and use other means to protect children. Hazards that cannot be kept secure may require parental training and guidance of the children for their own protection.

Types of Storage

Types of storage areas include:

- Enclosed storage areas—cabinets, closets, tool chest
- Horizontal surfaces—counter and bench tops, table tops, shelves
- Vertical surfaces—wall-mounted hangers or pegboards on which articles such as rakes, hoes, and shovels are hung for storage with easy access

- Floor surfaces—for freestanding, heavy, or bulky items; for machines that need to be permanently mounted to a strong base
- Ceiling surface—ceiling-mounted hangers from which items like a canoe or bicycle may be suspended
- Special storage areas—a place where items often used together are stored together, such as workshop or garden pest control materials

Each family must fit these things in available space, improve the efficiency of the available space, or add space for storage. If things begin to be too crowded, consider getting rid of seldom used items. Provide for flexible use of space and for efficient use of storage to avoid the cost of overbuilding storage structures.

The Garage

The garage will probably be important not only for the car, but for the variety of tools and power equipment associated with living on a few acres. The garage needs to be handy, large enough, safe, and secure. The best choice probably is an enclosed structure that holds one or two cars and also some of the garden and yard equipment you need.

The standard 20-by-20-foot attached garage is too small for anything other than two vehicles. For most homes on small acreages, a well-planned attached garage would be convenient and economical. Two to 4 feet added to the width and 8 to 10 feet added to the length or depth can provide for storage and work areas and still leave room for cars.

For example, a garage 24 feet wide and 28 feet deep would include useful work and storage areas, as well as ample room for driving in and out. This added space is less expensive than building the same space in a separate structure since you can use the same door, lighting, drains, and the like.

Ideally, though, you should have a plan for where you will store everything you want to put in the garage. Consider a work area, shelving, and convenient access to and adequate storage for the automobile.

A garage attached to your house has many advantages. It is convenient, secure, and costs less to construct at first. The convenience to the house has advantages such as easy access, maximum security, and economy if well planned when you are building a new house.

Frequently homeowners find they need a separate garage workshop where no garage now exists, or because the house has only a single-car attached garage. If the small acreage operation requires a limited amount of field machinery, it may be practical to design and build a combination garage-shop and machine shed. Considerations for the location of such a structure include:

- Storing the family cars
- Convenience from the driveway, the highway, and the fields
- Distance from the house because of security, convenience, frequent access to stored items, and exposure to the weather
- Heating and insulating the workshop area
- Storage of fuel for tractors or other power equipment
- Nearness of electric power
- Drainage and protection from wind
- Storing boats, camping trailers, and snowmobiles

If you store family cars in a separate building along with bikes, wagons, and other toys, the structure should be within 50 feet of the house. If it is primarily for a truck, field machinery, and garden tractors, place it farther away, approximately 75 to 100 feet from the house.

Drainage

Pick a well-drained site for year-round, trouble-free access in all kinds of weather. If good drainage does not exist naturally, it will generally pay to fill the area to provide drainage. The driveway and walkway leading to the building should be well graveled or preferably hard surfaced, with concrete or hot-mix blacktop.

For a heated workshop, select a structural system that is easy to insulate. Also provide an interior liner that is fire-resistant and easy to maintain. Plywood is a good surface to attach fasteners for storage shelves and cabinets, or for other items that are convenient to store on walls. Cover the insulation to protect it against mechanical damage and against flammability if one of the expanded plastics in board form is being used.

The floor of the garage-workshop area should be 6 to 8 inches above the surrounding grade and sloped to drain either to a floor drain or toward the garage door. A slope of $1/8$ inch per foot is usually adequate to provide good drainage.

A garage door should be at least 9 feet wide and 7 feet high. Field machinery may require wider and higher doors depending on the type of equipment used.

Heating with wood may be practical on many small acreages—especially if you have your own woodlot or there is an adequate supply of low-cost wood nearby. Allow plenty of room for the wood heater and install it with adequate clearance. Make sure the connection to an all-fuel chimney is done in accordance with recommended practices. Do not store any combustible materials near the heater.

Store the wood you plan to burn for nearly a year before use, so it is thoroughly dry. This means storage for wood should be large enough for a two-year supply. Obtain circulars that explain woodcutting, storing, and use for heating.

Store flammable liquids and gases in approved containers where there is no likelihood of flames or sparks. Some liquids produce explosive vapors, which if allowed to accumulate could create a hazard. Obtain circulars from your local Extension service that describe recommended safe storage and use of the proper type of fire extinguishers.

The Storage Shed

If you bought a property with a garage already built that is too small, it will be more economical to build or buy a low-cost storage shed. Prebuilt sheds are easy to erect, although they may not provide as much security as an attached garage.

These buildings come in metal or wood. They vary in size from about 6 by 8 feet to approximately half the size of a single-car garage. They can be built with no floor or placed on a concrete slab. Their most practical use is for storing yard tools and equipment.

Check the strength and durability of the doors. In prebuilt sheds, it's harder to store things that need shelves or wall hangers. Generally, these buildings aren't wired for electricity, so after-dark use is not convenient.

The Basement

Many one-story homes have generous basement areas, a portion of which could be used for a workshop. There are some disadvantages to this. One is the dust, especially if you do woodworking with power saws. It may be more practical to locate power saws in the garage.

Also, it is not convenient to store outdoor items in a basement unless it has a walk-out exit. Check the size of the door. The stairway should be convenient to the back door and planned so there is a straight approach to the stairway as you enter the house. The landing between the basement door and the outside door should be at least 4 feet long and planned so door swings do not conflict with each other or restrict movement to the basement. There also should be adequate space at the foot of the stairs.

Stairways to the basement should be at least $3\frac{1}{2}$ feet wide. Provide a handrail for safety and light switches and lights to illuminate the stairs. Be sure the stairs are well built, with a rise and run that provides for easy movement up and down. A riser of 7½ inches and a tread of 10 inches makes for a comfortable stairway.

Food Storage

SPACE FOR CANNED GOODS. A 5½- to 6-inch vertical clearance between shelves will hold Number 2 and Number 1½ metal cans, but glass pint jars need a 6-inch clearance. Allow a 3½-inch depth for a single glass pint jar or Number 2 or 1½ can, or 6½ to 7 inches if placed two deep. A 7½- to 8-inch vertical shelf clearance will hold quart glass jars. Allow a 4½- to 5-inch depth for a single quart jar or 9- to 9½-inch depth for two jars.

These dimensions allow for top and side clearance of the cans or jars. Storage space needed for canned goods varies, so plan space allowances according to your family's needs.

Canned goods can be stored in a basement or pantry. Temperatures of about 50 to 60 degrees Fahrenheit are desirable, but should not go below freezing or higher than about 110 degrees.

FROZEN FOODS. Use a freezer for storing foods longer than three weeks. Either chest or upright-type freezers are heavy when full, so be sure the floor will support them. Do not place the freezer where the room temperature goes down to freezing or below, and avoid locations where the sun or a heat source will cause the motor to run often. Avoid locations with high humidity. Try to have a counter or table nearby to use while loading or unloading food packages.

Frozen foods can also go in a rented, commercial frozen food locker, and packages can then be brought to the house for brief storage prior to use.

FRUITS AND VEGETABLES. Fruits such as apples and pears, and vegetables such as potatoes and onions, can be stored in a small basement room. This room should be on the north side of the house, walled off and well insulated from the rest of the basement. Properly built, such a room can also serve as a disaster shelter, unless you have a separate shelter room located on the side of the house from which storms approach. A window is not necessary, but a vent to the outside with a damper can maintain the desired temperature and get rid of vegetable odors.

The humidity should generally be moderately moist—moist enough to avoid drying the produce and dry enough to avert spoilage. Natural air movement or use of a fan in the storage room also helps prevent mold growth.

Various fruits and vegetables require different temperatures, humidity, ventilation, and other conditions for optimum storage. Your county Cooperative Extension office will have information about methods and conditions of storing produce in your geographic area.

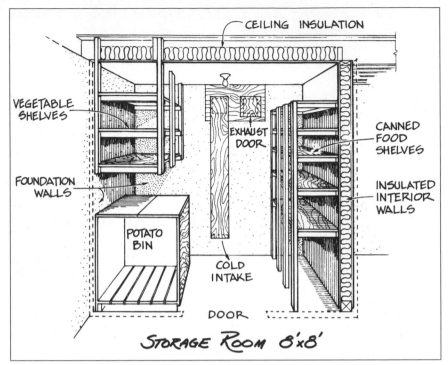

CEILING INSULATION

VEGETABLE SHELVES

EXHAUST DOOR

CANNED FOOD SHELVES

FOUNDATION WALLS

INSULATED INTERIOR WALLS

POTATO BIN

COLD INTAKE

DOOR

STORAGE ROOM 8'x8'

A small food storage room, located on the north side of the house, should be walled off and well insulated from the rest of the basement. Natural air movement or use of a fan will keep produce dry enough to avoid spoilage.

Determine your needs for produce storage and provide bins, shelves, and hangers for mesh bags. Check the produce regularly and dispose of items starting to spoil.

Properly constructed outdoor cellars are excellent for storing many vegetables, especially root crops. They also can serve as a storm or other disaster-type shelter. Built below ground, they offer a uniform temperature but must be strong enough to support the weight of several feet of soil above and soil pressure on the sidewalls.

Build where there is likely to be no water problem, or else construct a tile drainage system that will keep water out.

It is likely a root cellar would be more economical to incorporate into a new house if you are building with a basement.

Lighting

Lighting is of two types: general lighting and task lighting. A lighting intensity of 10 foot-candles is adequate for general overall lighting. Task lighting varying from 20 to 50 foot-candles is recommended at the workbench, depending on the detail of the tasks at hand. Some work may require more light.

Ceiling lights commonly provide general lighting. A reflector above the lightbulb will help direct the light downward. Light-colored walls and ceiling help reflect light and aid in general lighting. Place lights about 10 feet apart and about 6 feet from the side at a ceiling height of 9 to 10 feet. Hundred-watt bulbs will generally provide necessary overall lighting.

Over the workbench, you need a higher-intensity light. If you use incandescent lights, install two to avoid shadows on your work. Two 150-watt lamps placed 4 feet above the workbench and 4 to 5 feet apart will provide about 50 foot-candles on a 3-foot-high workbench. Two 40-watt fluorescent tubes will provide about the same light level. A fluorescent fixture requires a higher initial outlay, but the cost of operation is less.

It may be desirable to plan for portable lights in case the workbench is to be moved in the workshop area. When a close-up light is needed at a saw or drill, a portable lamp holder with a 60-watt bulb will generally do the job.

Plan convenience outlets for flexibility of use. Outlets should be no more than 10 feet apart. This means that you won't have to reach more than 5 feet along the wall to any convenience outlet. Circuits should be properly fused. Provide 15-ampere service for lighting outlets, and 20-ampere service for electrical tools. Be sure outlets are properly grounded and the equipment you use is fitted with the three-prong plug.

In a large garage-workshop it may be desirable to provide ceiling drop cord outlets to avoid long extension cords from wall outlets. If more powerful motors are used, it may be best to up the voltage of your outlets. The electrical service entrance must provide for this, of course.

Light the area above the entrance doors. Place a 100-watt bulb 10 feet above each garage door. Install switches in the house as well as in the garage. For added convenience and security, install a radio-controlled garage door opener.

Costs

Improving existing storage space will cost less than adding storage space. If you can do the work yourself, installing pegboard and hangers on walls may range from $10 to $50. Adding shelves will probably cost more, and adding closets and cabinets will be the most expensive.

Costs of improvements or additions can range from a few dollars to several hundred. Small costs for occasional hardware and material purchases can often be handled as part of the weekly or monthly family spending.

Before you proceed with a moderate or large-sized job, determine probable costs for the improvements or addition of workspace and related storage space. To do this, first get your plan on paper. This gives you a basis for estimating the cost of construction and materials if you do the job entirely yourself or subcontract some parts of it. If you plan to have the work done by others, the plan will be a basis for obtaining bids.

Families may assume the financing from their savings. Or they may seek financing from outside sources. Conventional lenders include savings and loan associations and banks.

Landscaping: Get Help, Plan Carefully

You have almost unlimited opportunities when landscaping. If you plan well, you can find room for most outdoor activities, from gardening to small farming ventures.

Visibly enhance the house and outbuildings with plants. It should look good and be easy to maintain. Plan it to be useful, beautiful, and to add value to the land.

Although you can make your landscaping plan on your own, think about consulting with a landscape architect or landscape designer. You could hire a consultant for a preliminary study and, combined with reading and self-study, complete the plan yourself. There is a great deal of garden literature about small properties.

Each residential planting plan can be unique. No two houses, sites, or families are exactly alike. It follows that no single landscape plan will fit all properties or answer the requirements of all families.

Try to enhance the architectural style and lines of the house. The building materials, colors, and entryways can offer clues and suggest landscaping ideas for plants and structures. But most important, the planting should be designed for the people who will live there and use the land.

First, gather information on the conditions of the site and the needs of the people who will use the land. Consider the amount of time and expense family members want to devote to grounds maintenance.

A limited budget should not prevent you from developing a landscape plan. Larger projects or extensive plantings can be phased in and budgeted for over a period of years. This is why a landscape plan is an essential first step. It permits all parts of the total landscape to be fitted together at later times, like parts to a puzzle. When carefully planned, the finished landscape will be a complete and pleasing picture, rather than a jumble of plants and accessories unrelated to each other.

Study the Site

The first step is the orderly and logical recording of conditions and facts on the buildings and land. You must overcome or modify the restrictions and enhance the assets. List the conditions, natural or man-made, that have immediate or potential effects on the property. These can include anything you hear, see, smell, or feel.

The size and shape of the land, direction of the sun, winds, and views all present restrictions or opportunities to landscaping a small acreage. Get the feel for the land to understand what the site has to offer. Success or failure in producing a functional landscape plan often depends on how well you or a designer understand the site's characteristics.

Organize the land and outdoor spaces to suit you. The landscape should look good twelve months of the year, not only in the spring and summer. Plan the buildings, plants, and structures to strengthen each other. No amount of planting can overcome the lack of good organization.

You can't eliminate the natural forces of sunlight, rainfall, winds, frosts, and temperature. However, a landscape can modify their effects. Broad categories to consider are: climate, topography, land, soil, vegetation, house, utilities, and the community. Locate, describe, and evaluate these conditions. Judge each condition. Is the condition useful? Is it good or bad?

Climate and weather affect outdoor activities more than any other factors. Plants or structures can be used to create shade, trap heat, redirect or slow wind movement. Minimum temperatures determine the kind of plants that will grow. Temperatures also determine the range of outdoor work, gardening, and recreation activities. Rainfall, frost-free periods, and wind direction all affect landscaping.

The changing direction of the sun from winter to summer creates a whole different set of sun and shade patterns on the land and buildings. Knowing where there is sun or shade at different hours and seasons helps solve problems for plants, gardens, and outdoor activity areas.

Road noise, traffic, glare, and streetlights also affect the landscape. Consider how the pattern of street- and auto headlights on the windows and outdoor areas may impose a set of restrictions on benefits that could be modified with plants or structures.

Consider the topography. Is the site sloping, or rolling? Will the planned activities work on the existing grade? Perhaps the activity should be changed. An alternative would be to modify the site by grading if the hobby or activity has a high priority.

Aesthetically, sloping or rolling land is more dynamic and has more advantages for the design of the house and landscape. A level site has neutral and minor landscape interests. On a flat site, try bold colors and exotic materials. Level areas have less protection from the wind, so you might need more climate control elements such as trees (which act as windbreaks) or structures.

Drainage and grading are closely related problems. If you intend to grow plants, you might need to create drainage or grading. Do not allow runoff to drain over the top of the property. Seek help with drainage or grading problems through an engineer, landscape architect, the Cooperative Extension, or a landscape contractor.

Soil information is needed for the site analysis to determine what plants (shrubs, flowers, vegetables, fruit) will grow best on the land, or if changes must be made. A soil test can be obtained through the county Extension Service. Results of the test will indicate the soil's lime requirement, its fertility status, and if corrective or maintenance fertilizers are needed.

Knowing the soil helps you forecast the potential for growth, development, and success of a landscape planting, horticultural venture, or hobby. Look at the native plants. These enable you to "read" the landscape. They provide clues to the local environment and soil conditions. Trees, such as red maple and sour gum, and shrubs, such as arrowwood viburnum or red-stemmed dogwood, can indicate wet or poorly drained soils.

Identify and evaluate why the existing plants are growing on the site. Are they worthwhile? Do they add to the landscape? If they are removed, will this adversely affect or improve the area? Some native plants may be an asset, some a liability.

The House Plan

The house exerts a strong influence on landscape design. The house plan will affect the relation of the house to the outdoor areas and gardens. Rarely are the house and garden designed together. A door from the kitchen or dining room is the logical place for outdoor cooking and eating. A door from the living room to an outdoor patio is a logical place to sit or entertain. A door from a bedroom could lead to a private garden.

Assess the views from inside the house looking out and from outside looking in. Should they be screened, hidden, or used? To be able to look into a neighbor's attractive landscape is like owning the land without the taxation.

Consider the impact of movement of people and vehicles about the place. Dimensions for outdoor-use areas—such as walks, steps, patios, and furniture—are larger than those for indoor use. Walks should be wide enough for two people side by side, at least 4 feet wide.

The house tends to dominate on a residential property and be less dominant on an acreage. Most houses are geometric, and the land around them can be developed with the same geometric or formal pattern. A formal plan is the easiest and safest for a landscape, especially for small areas. On larger properties, the informal free-form design works best. You can design a formal landscape around your house, and, beyond it, the plantings can give way to a more informal arrangement.

Trained landscape designers know how to integrate the land and what you will do on it in a pleasing way. As the owner, you might not be able to express why you like what you see, but you will know whether it pleases you. On the plan, mark the locations of utility lines for water, sewer, electrical, and telephone. The location of meters, height of wires, sewer clean-outs, and tile lines all influence the placement of plants and gardens; so do walks, driveways, and property easements.

Consider how people drive and walk on your property. Landscaping can enhance what people will feel, see, or experience as they come onto the land. Are there enough lights for entering at night? Are the entrances visible?

Zoning and building regulations will control some of what you do. These rules can protect or restrict the building of a swimming pool, privacy fences, or boundary line plantings. Do not consider zoning regulations a hard-and-fast contract between local government and property owners. Regulations and conditions have to change with the times, and the appeals process continually tests and alters them.

Nothing to Hide

One of the most inappropriate landscape approaches in use for the last century is the foundation planting. This technique was based on a nearly abandoned house form. Foundations of many of today's houses don't show. There is generally nothing to hide—and yet many owners continue to plant bushes and flowers around the base of the house as if the foundation is sticking up.

Consider this alternative: Forget the foundation planting concept and treat your entire property—house and land—as one living environment.

Plan for People

You don't have to force a plan onto the site. Think about the people who will use it. Thoroughly know what you need, what you want, and what you can afford. The number in the family, how old the members are, and what they like to do all reflect your landscaping. Children's play areas and outdoor cooking, eating, and entertainment areas could be important. Choose the best locations for hobbies and garden ventures for fun or profit. Consider also how much space you need for storage, parking, roadways, service areas, animals, or house pets.

Few people can afford to complete a landscape all at once. A loan can finish the job sooner and might be a good idea, if you can justify the extra cost in the immediate use and enjoyment, plus the property's added value. Realize that once you begin developing the landscape plan, the realities of cost or difficulties in the land itself may alter your plans.

The Plan Itself

Using all of the information you have gathered, put together your plan—or wait while the professional you hire does so. It will include written steps and diagrams. You can hire a landscape architect or devise a plan using some of the computer software now available. The plan begins with a bird's-eye view of your property (on paper or computer). Over the basic drawing of existing features (like the house, driveway, trees, and boundary lines), you can add (with tracing paper or on computer) physical elements that affect your plan, such as shadow patterns in different seasons, good and bad views, slopes, drainage areas, and the direction of the wind. On your plan, you can note possible solutions to these challenges.

Then, figure the interrelationship of the outdoor-use areas in terms of rough size, abstract form, and sizes by drawing "bubble diagrams." You can do this by using another sheet of tracing paper, if you are planning on paper. The bubble diagram is a way to try out ideas on areas of the land. For example, if you want to consider a strawberry patch outside the back kitchen door, and a bed of herbs next to it, with a deck between that area and a grove of small trees, each of these features would be represented by an amorphous circle on the bird's-eye view. Inside the bubble you label what you plan to try growing there.

These diagrams help you evaluate the outdoors and how it relates to the house, driveway, and walkways. These bubble diagrams (or functional diagrams, as they also are called) help you check whether or not each area fits your needs or desires. They help in evaluating the impact of the use area on the soil, vegetation, neighbors, or community.

The outdoor eating and cooking area, or the patio, should be in an area for easy transfer of food and dishes from the kitchen. Storage facilities for patio furniture, garden tools, or play equipment need to be nearby. You can decide if it is worth modifying the land for what you want to do with it, or modifying what you try to do to preserve the land as it is. For example, you might grade a slope to build the patio

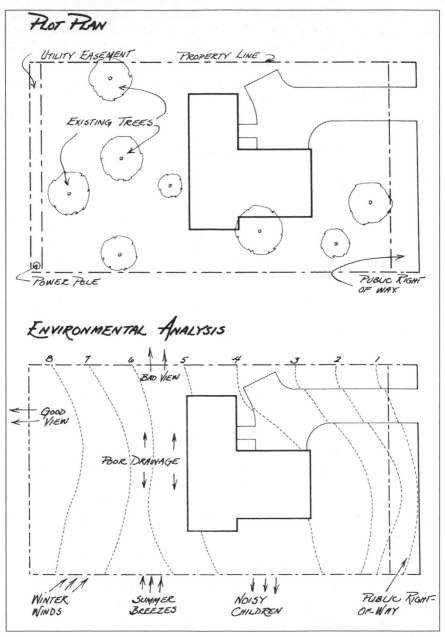

PLOT PLAN

UTILITY EASEMENT   PROPERTY LINE

EXISTING TREES

POWER POLE

PUBLIC RIGHT
OF WAY

ENVIRONMENTAL ANALYSIS

8   7   6   5   4   3   2   1

BAD VIEW

GOOD
VIEW

POOR DRAINAGE

WINTER
WINDS

SUMMER
BREEZES

NOISY
CHILDREN

PUBLIC RIGHT-
OF-WAY

Professional plans can help you evaluate your landscaping needs while taking into account certain environmental challenges such as drainage areas, undesirable views, and shade patterns in different seasons.

away from the kitchen door in order to preserve a tree that shades the kitchen window.

Outdoor Rooms

FUNCTIONAL DIAGRAM. You now have established outdoor "rooms" that can be divided, separated, screened, or connected with plants or structures. The functional diagram is the germination of your ideas.

The next step to the design process is development of a preliminary plan to locate the plants and structures used in the design. These are the elements that, in an abstract way, create the outdoor rooms.

Indicate all plants and structures in terms of their width, height, length, and functional purpose. For example, a tree, awning, or arbor could provide shade. A privacy screen could be either a hedge, wall, or fence. In this phase, the approximate sizes are shown in an abstract fashion for the structures, paved surfaces, locations for plants, areas to be shaded, and changes of grade.

This form of preliminary plan gives a picture of the proposed design, not detailed enough to work from, but detailed enough to test the program. Examples and pictures from books and magazines can supplement the design.

THE BLUEPRINT. The final design is the master plan. This plan is the blueprint showing to scale the exact location for all structures, pavements, and plants, plus their names and/or building material. At this point, you or your designer will determine the arrangement and form of the plants and structures.

Effective landscape design is not as simple as it may sound. You must solve certain environmental and space design problems by considering the functional spectrum of plants and applying the design principles of simplicity, balance, scale, sequence, and focalization.

The time to begin a maintenance program for the yard is when the landscape is first designed. With careful thought, maintenance does not have to be drudgery. Many ambitious landscape designs can make unnecessary work for the owner if planned for maximum effect rather than the minimum maintenance most people want.

As the landscaped area begins to develop, change, and mature, so does the family. Anticipate their requirements in a landscape design that should be flexible enough to adapt to your family's changing needs.

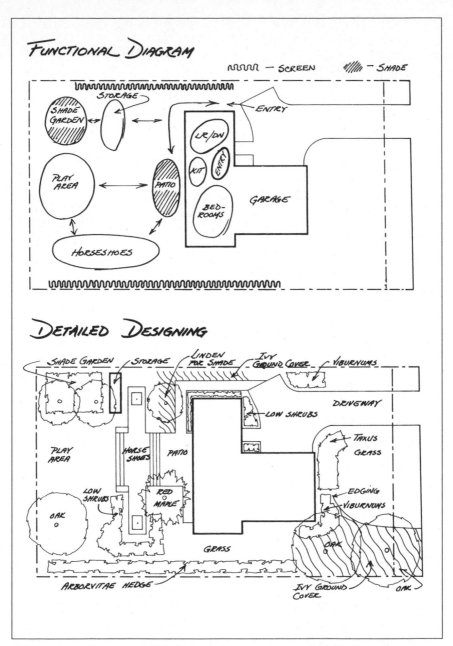

Functional Diagram

ᴧᴧᴧᴧ — SCREEN   ▨▨ — SHADE

STORAGE

SHADE GARDEN

ENTRY

LR/DN

KIT   ENTRY

PLAY AREA

PATIO

GARAGE

BED-ROOMS

HORSESHOES

Detailed Designing

SHADE GARDEN   STORAGE   LINDEN FOR SHADE   IVY GROUND COVER   VIBURNUMS

DRIVEWAY

LOW SHRUBS

TAXUS
GRASS

PLAY AREA

HORSE SHOES   PATIO

LOW SHRUBS

RED MAPLE

EDGING
VIBURNUMS

OAK

GRASS

OAK

ARBORVITAE HEDGE

IVY GROUND COVER

OAK

After you have determined what plants and structures you would like around your home, a landscape designer can turn your functional diagram (top) into a master plan (bottom) that will include plant names and the proper scale of all buildings.

Land Improvements—
What You Need to Know

Chances are you will not find the ideal country place with all the features you desire. Improvements such as water supply systems, drainage systems, windbreaks, fences, and roads usually cost a lot of money. It's important that you be able to recognize the value of existing improvements and assess the feasibility and cost of improving submarginal land to meet your needs.

If you are looking for a country place with the desired improvements already established, learn to evaluate the quality of the improvements. Are the fences in good repair? Is the dam that creates the farm pond sound? Is the horse pasture well grassed? Are drainage or irrigation systems functional? Are there signs of serious erosion?

If you intend to start from scratch on an undeveloped piece of land, you need to be especially cautious in assessing the potentials for improvements. This is particularly true if you plan to move to a region with which you are unfamiliar. Never buy land sight unseen. The climate and other conditions could be drastically different than you expect. The following sections are intended to provide basic information to help you assess the quality of existing improvements and to evaluate the need, feasibility, and cost of the improvements you may want.

Water Supply

If you intend to establish a country home on an undeveloped tract, your first concern should be a water supply. The availability of water varies considerably in different parts of the country. In areas of high precipitation, water-bearing aquifers frequently lie near the surface and wells can be shallow, dependable, and relatively inexpensive.

In more arid regions, this may not be the case. Water is a scarce commodity in much of the American West. Many western states are concerned about receding water tables. You need permits to construct all wells, and you should not count on getting them.

Even if you have the required permit, the cost of drilling a well may be prohibitive. Depth to water and the kind of geological materials between the surface and the water generally govern the cost. A driller can make a new well for between $6 per foot (in Arkansas) to hundreds of dollars (in Arizona) per foot. In many areas, wells reach reliable water sources in less than 100 feet. In others, the closest water may be at 500 to 600 feet. Depending on the depth to water, then, and the difficulties drilling to it (whether the driller must go through rock, for example), wells can cost from $1,000 to easily $100,000.

Another concern in developing wells is water quality. In some areas groundwater is so saline that it is unfit for human or livestock use. In mountainous areas, groundwater is frequently polluted by poorly designed waste disposal systems installed upslope.

A good source of information about the cost and feasibility of wells is the local well driller. He usually maintains drilling logs and is familiar with local conditions. The state agency that issues well permits may also be able to supply useful information. It may even be worth your while to hire a consulting geologist if you have serious doubts about the availability of water.

Undeveloped springs are sometimes a good source for both domestic and livestock water. Look for seepy areas that might indicate a near-surface release from a water-bearing soil layer. Even if ponded water is not evident, you may be able to develop a good water source by installing a collecting wall and gravel-packed perforated pipe. With proper storage facilities a flow of a gallon or two per minute should meet your needs.

Be sure to check the quality of water from a spring. Your county Extension agent can usually arrange water-quality testing for a small fee. Developed springs should be fenced to prevent pollution from livestock.

Drainage Systems

Excessive soil water can be a serious limitation to most proposed land uses. Severe drainage problems may go unnoticed by the inexperienced. Land that is dry during the summer months may turn into a marsh during spring rainfall and snowmelt.

Learn to associate certain kinds of plants with soil moisture conditions. The appearance of sedges, rushes, or other water-loving plants is reason to suspect a high water table. In areas of limy or alkaline soils, presence of salt crystals on the surface also indicates a high water table. These are brought to the surface in solution by excess water and precipitate out as the water evaporates.

Land with a high water table can sometimes be converted to highly productive uses by installing a drainage system. But be forewarned: With today's federal and state wetlands laws, there is a good chance it's illegal to drain wetlands. Planning officials at your local municipal offices can direct you to the local land-use regulations. Even if it's legal, you will need expert advice to determine the feasibility of drainage. Also, the cost of drainage is highly variable and must be determined through onsite study. You will need to compare the cost with the benefits you expect to receive.

In areas of high rainfall, a good surface drainage system may be necessary to carry off water from high-intensity storms. If you are considering purchasing a small unit subdivided from a larger farm or ranch, make sure your lot is not located near the outlet of a terrace system or in an unprotected floodplain. Structures to protect you and your land may be expensive, impractical, or not permitted because of local restrictions.

Pasture Improvement

Overestimating the carrying capacity of land for livestock is perhaps the most common mistake new country dwellers make. A saddle horse will eat 30 to 40 pounds of forage a day. A typical nonirrigated pasture in many parts of the country can produce only 500 to 1,000 pounds of usable forage per acre per year. This means that without supplemental feed, you need 30 to 50 acres of this land to pasture one horse for a year without degradation of the forage and soil resources.

Pasture-carrying capacities can be significantly increased by irrigation, rotation grazing systems, fertilization, or establishment of better species. But even with an improved pasture system, 1 acre of land will support a horse for only about four months.

Overgrazed pastures soon become barren or weed-infested areas that are erosion and health hazards. Local ordinances may require you to install erosion control measures and curtail grazing on such areas.

Feeding hay or grain to your livestock, or leasing additional pasture to sustain them through the year, can be a significant unexpected expense.

Windbreaks

A mature windbreak of trees and shrubs around the homestead or along field boundaries is a valuable improvement. It may take twenty years to grow an effective windbreak, but once established it will reduce fuel costs, protect property from high winds, reduce soil blowing, keep snow from drifting against buildings, attract wildlife, provide shelter for livestock, and add beauty to the landscape.

To be successful, windbreaks must be properly designed and consist of adapted species. The most effective windbreaks contain three to five rows, with the lower-growing species on the windward side and the tallest species to leeward. One row of trees will not do the job. Choose evergreen varieties so that the trees will do their job even in the winter. Place individual plants in the rows. Carefully figure the distance between the windbreak and the house, so that windblown snowflakes do not bury it.

Windbreaks are relatively inexpensive to establish. For large windbreaks, buy seedlings that are two to three years old from commercial nurseries, forestry agencies, or other conservation agencies. The price varies depending on the species and region, but, for example, in Missouri, the State Forest Nursery sells bundles of twenty-five seedlings that cost between $3 and $18.

Smaller windbreaks can shelter the side of the house that bears the brunt of the weather. For this smaller grouping, you should buy more mature trees and expect to pay between $20 and $100 per tree at a private nursery. Plant by hand or with planting machines, which you can rent in some areas. In areas of high winds, hammer a cedar shingle into the ground on the windward side of each tree to protect it during establishment.

To control weeds, cultivate the soil between the rows for the first few years after planting. In areas of undependable precipitation you might consider a drip irrigation system. These systems use very small amounts of water and are used only for the first two or three years to assure establishment.

Seek technical assistance for windbreak planting from conservation districts, county Cooperative Extension offices, the U.S. Natural Resource Conservation Service, and state forestry agencies.

Farm Ponds

A farm pond is an attractive improvement to any country unit. It looks good, provides livestock with water, and offers recreation, fish and wildlife habitat, and fire protection. A good pond site is one where the largest volume of water can be stored with the least amount of fill in a dam. Look for a place where the valley is narrow at the dam site and the reservoir area is wide and flat.

The size of the watershed above the site must be large enough to supply the water needed to fill the pond, but if it is too large it may be expensive or impractical to construct a dam and spillway to handle peak flows. Evaluate the extent of active erosion on the watershed above the pond. Too much erosion can fill your pond with sediment in a few years. It may be wise to delay building a dam until the watershed is stabilized with vegetation.

You will need to determine the suitability of the soils before building a pond. Soils with a high clay content are generally best, since they are relatively impermeable. Sandy or gravelly soils do not hold water and are unsuited for both dam construction and reservoir area.

It is necessary to clear the pond area of existing vegetation before beginning construction. This eliminates safety hazards and the possibility that the decomposition of plants may cause the dam fill to become unstable.

The depth of water is important. Shallow ponds produce excessive aquatic vegetation that is detrimental to most uses. Specific water depths are required for various species of fish, especially where live water does not move continuously through the pond. Do not construct ponds near feedlots, sewage disposal fields, mine dumps, or other pollution sources.

If you cannot locate a good site on a natural drainage channel, you might investigate the possibility of an off-channel pond. This type of pond depends on an alternate water source, such as a spring or diversion from a nearby stream. Off-channel ponds have some advantages over the in-channel type. You can control the water supply and usually eliminate sediment problems.

Test the water quality before you build a pond, especially if you intend to stock it with fish. Most fish species have definite tolerance limits to toxic elements, pH levels, and water temperatures.

The cost of building a pond may vary considerably, depending on size of the dam and the complexity of the site. Many states have laws that regulate the use of water and that specify acceptable criteria for dam design. Before you start building a pond, find out what laws apply.

You will need engineering assistance to help you select a proper site and to prepare plans and specifications for the dam. You can obtain this kind of help from your local Natural Resource Conservation Service district, or you may want to employ a private engineer.

Fences

Fences mark property boundaries, control livestock movement, regulate access by people, and protect property. They may even add to aesthetic value.

The most common fences in use today are the barbed wire fence and woven wire fence. (Horse owners should be aware that barbed wire fences can cause injury to horses, especially around corrals or small pastures.) There are many types of wooden fences, but their purpose is now primarily ornamental. While attractive, they are very expensive and much less effective in controlling livestock.

Fences represent a significant investment. A good barbed wire fence will cost about 83 cents a foot installed. Woven wire fences cost more, about $1.07 per foot. A barbed wire or smooth wire fence should have at least three strands of 12$\frac{1}{2}$-gauge galvanized wire with class-2 zinc coating. Additional strands may be needed to confine small livestock such as sheep, goats, or ponies. Fence posts may be wooden or steel and should be spaced no more than 20 feet apart.

Wooden posts should be at least 3 inches in diameter and 6 feet long. Place them at least 18 inches into the ground. Corner and brace posts should be at least 5 inches in diameter, 8 feet long, and placed at least 3 feet into the ground and anchored.

Drive steel posts into the ground so that the top of the anchor plate is level with the soil surface. They should be long enough to allow for a fence height of at least 42 inches.

Wooden posts made from cedar, juniper, osage orange, catalpa, black locust, or redwood contain natural preservatives. Most other wooden posts should be treated with a commercial preservative. Standards and specifications for fences are available at local soil conservation districts.

Electric fences, which are charged-wire fences, can keep a pushy animal from trying to break loose. Six-strand high tensile fences cost about $1 per foot, installed.

Roadways

Do not underestimate the importance of a good access road to your country house. If improperly designed, it may cause you considerable inconvenience and be a continuing expense. If you plan to build a new roadway from a public road to your house, you need information about soil properties, topography, and drainage patterns. You also need to evaluate how well the proposed roadway will function under the most severe weather conditions.

Soils with high clay content become very slippery when wet. Sandy soils frequently lack stability and erode easily. Most roadways require surfacing with gravel or other suitable material. Place gravel about 6 inches deep on the surface. The common 10-foot-wide tread width will require about 20 cubic yards of gravel per 100 feet of road.

Gravel costs vary and depend largely on delivery distance. Gravel costs about $25 to $60 per cubic yard, depending on the type of gravel, not counting delivery and spreading costs.

Slope stability is an important consideration, especially in steep country. Roads cut into steep hillsides can fail completely if soils are unstable or have moisture moving through the soil profile. Try to plan your road to avoid steep grades. If possible, keep the roadway grade below 6 percent. Steep roads become extremely hazardous when covered with ice or snow.

Improper drainage causes most unpaved roadway problems. The road should be sloped laterally to prevent water from running down the road surface. Small graded ridges called water bars are helpful in getting water out of ruts and into a planned drainage system.

Natural drainageway crossings are especially critical. Even the smallest gully may become a torrent during a high-intensity storm. Most drainage crossings require a culvert, bridge, or grade dip to keep the water in its natural channel without damaging the road. Be sure to design bridges and culverts to be large enough, and install them at the proper grade, as sediment and debris reduce the culverts' carrying capacity.

New roads can erode the land around them, especially where cut and fill slopes remain exposed. Plant these areas to stabilize the slopes as soon as you can. You may also want to plant trees or shrubs to screen unsightly areas.

If your road project is complex you might need an engineer or surveyor to help you with planning and design. Local road departments are good sources. For help planting after you build the road, contact soil conservationists, county Cooperative Extension agents, and forestry agencies.

Conservation Measures

As you make improvements, take steps to protect soil and water resources, improve productivity of the land, and create better environmental values. On irrigated lands, conservation measures include such practices as land leveling, ditch lining, drainage, pasture establishment, and structures for water control.

In nonirrigated farming regions, where erosion from both wind and water are prevalent, common improvement practices include terrace systems, diversions, grassed waterways, grade stabilization structures, windbreaks, and small dams. A number of conservation practices are designed to improve range and woodlands. These include livestock water developments, brush control, proper grazing management, and various tree management practices.

Special improvement practices to create benefits for wildlife or establish facilities for recreation are also considered conservation measures. Several government agencies have programs designed to help the rural landowner with planning, financing, and carrying out land improvement measures relating to conservation.

If you move to the country, get acquainted with your local U.S. Natural Resource Conservation Service district. It can give you the technical assistance needed to plan and lay out conservation practices.

If you need financing to establish land improvement practices, contact your county representative of USDA's Farmers Home Administration. This agency makes low-interest loans for certain rural land improvements.

The U.S. Forest Service, through its state and private forestry program, provides excellent assistance in forest management and related activities. Many state forestry agencies provide similar assistance.

Your Farmstead Buildings

The type of homestead you have will determine the kind of buildings you need or can get by with, ranging from windbreaks to barns. A suburban estate or rural acreage within organized communities requires that structures be somewhat conventional. They will probably have to meet building codes, and the aesthetic value may be important. Since the owner will have a source of income, the cash cost may be a minor consideration.

On the other hand, homesteaders far removed from codes, jobs, power lines, and neighbors will be primarily interested in low cash outlay and the structure's utility. The remote homesteader has much more latitude in design, and often will use different materials. In fact, the design depends on the choice of materials at hand. The labor must be done by the homesteader based on sound building principles, with the emphasis more on permanence than beauty.

The financial situation of many homesteaders will be somewhere between the two examples. Most designs, then, will be a blending of features to accomplish the most desirable compromise of low cost, convenience, function, permanence, and beauty.

In all cases the structure will be a capital asset, and its effect on the overall value of the estate is important. Even though the structures may be planned for your sole use, consider what value different designs will have on your property value should you sell the property. In some cases the estate's increased value may be enough reason for a construction project, since many homesteads have been created only to be sold at a tidy profit.

The Building Site

For buildings to fulfill their purpose, they should be built according to an overall plan. Although all good homesteads have general qualities in common (buildings are convenient to road access and the house, for instance), differences in individual terrain make planning for each homestead a necessity.

The first consideration in locating the buildings will be freedom from surface water drainage. Buildings should be situated on high ground. When they must be located on slopes, create drainage ditches so that water doesn't enter the structures. Each animal shed should be located where water can't accumulate; if this is not possible, you must build a mound of some sort. In all cases, drainage from the roof should run away from the building. Your plan should place buildings away from drainage channels.

Take care to see that contractors or other construction workers do not leave a depression around building foundations, or that settling of fill does not result in low spots that are apt to collect water adjacent to structures. You also will need to see that future buildings do not block drainageways from your initial ones.

Observe the direction of local winds when locating structures, especially in northern climates where drifting snow may pose a threat to animal housing. Use natural windbreaks, and consider plantings for this purpose. Don't let valuable land blow away, or allow your animals to become sick or unproductive. If your site is wooded, cut only those trees that need removal; but don't allow excessive shade on your garden if you expect it to be productive.

Many factors will influence the location and orientation of your buildings. Winter sun shining into animal enclosures will do much to keep them dry and aid the health and productivity of the animals. Study possible overflow from nearby streams that may become frozen or flooded. Many streams overflow in winter when ice restricts the flow in the main channel. Neighbors or old-timers in the area may be able to advise you on quirks of nature that may affect the building sites you pick.

Of course, your basic wants and needs will be paramount. Estimate the size garden you need, and the kind and number of animals you wish to care for. Consider every factor you can in creating your homestead design within the limitations imposed by the terrain.

Building Materials

Every building has two basic parts: the structural framework and the weather-protective material. A weatherproof surface material and a separate insulating layer generally protect the building from weather. Sometimes one material provides structural strength, weatherproofing, and insulation.

Depending on the structure's purpose, and the climate, more or less of each building component will combine to provide the environmental control you need. The properties of the materials determine how long the structure will last. For instance, adequate roof overhang to avoid wet walls can compensate for using a building material that deteriorates or rots readily when wet.

Many conventional structures use a wood frame of poles or studs to support the roof. Cover walls with wood or metal to cut down air exchange, and attach a nonstructural insulating material to the frame to prevent excessive heat loss. Add a waterproof material to the top of the structure to protect against rain and snow.

Often building materials can provide two or more required properties. Logs or sod, for instance, have structural strength, and can be cut to restrict airflow. They have good insulating value in the thicknesses commonly used and provide resistance to deterioration from moisture.

If a stout frame can be made, many materials will provide windproofing and insulation, especially if an adequate roof prevents major damage from moisture. Hay, straw, or moss can provide good weather protection and warmth.

Remember, the key to homestead living is to cut cash expenditures. You can do this if you use ingenuity in your own designs and rely on materials gathered locally at little or no cost.

Ventilation

A primary design feature in structures, especially for animal housing, is ventilation. Most animals can withstand cold, but not wet and drafty conditions. Animal hair gives good insulation unless it's ruffled by air currents or gets wet. Ventilation removes moisture from animal structures. Air enters the structure and is slightly warmed by the animals' body heat. This warmed air is able to hold more moisture in the form of water vapor than does cold air, so it absorbs the excess moisture in bedding and from the animals' breath.

Replacing this warm moist air inside the structure by cool, dry, outside air provides the drying effect of ventilation. The design of animal structures must provide for this essential air exchange. Besides its drying effect, ventilation provides fresh air for breathing and air exchange to carry away undesirable odors and air contaminants.

As the structure is made drier by ventilation, the inside environment becomes a poorer place for the increase and spread of disease germs or organisms that cause sickness in animals. In enclosed buildings, holes or slots are left in the structure to allow for air exchange.

For short periods when wind and temperature are severe, the holes can be closed. Open up the building when the storm is over so excessive moisture, odors, and air contaminants don't build up. Removing moisture from the structure by ventilation reduces a major threat to structures as well as to animals. Rot in wood is caused by microorganisms that can grow in the wood only when moisture is present. The rot stops when the wood dries out, but becomes active if the wood gets wet again.

The most common protection against moisture and wood decay is to place structures on concrete foundations above sources of ground moisture. Another way in moist locations is to use wood that has been treated with a material toxic to the growth of rot organisms. Most of these toxic materials resist insects that burrow in wood as well. Many also deter attack by rodents or larger animals.

With treated wood, cheaper designs can be used, since considerable strength is gained by building around treated posts planted in the ground. Buildings that might otherwise last only four or five years can last thirty to fifty if made with foundations of treated wood.

Sources of Plans

If you want a conventional animal shelter, or storage for crops or machinery, there are several good sources of plans. State universities maintain plan services with plans suited to the climate of that area. Building supply stores not only sell lumber, but have plans for structures. Plywood, lumber, and concrete trade associations will supply plans showing how to use their products economically and safely. Metal building manufacturers not only provide plans and materials, but also probably can suggest a reliable contractor in your area to erect the structure if you prefer. Check any farm magazine for its abundant information about commercial agricultural buildings.

If you use native materials found on or near your land for little or no cost, then you likely will have to create your own designs using the building principles discussed earlier. You can save money by cutting trees to take to the sawmill. Many small mills will saw your timber into rough lumber on a shares basis, allowing you to use more conventional materials and readily available plans at reduced cost.

Don't disregard dry straw, grass, or weeds as insulating material. Wood shavings or sawdust are good sources of inexpensive insulation. Most of these materials have about half the insulating value of commercially available insulations, so you will need about twice the normal thickness.

Some materials that you can provide for yourself, like ground sphagnum moss or coarsely ground peaty materials, insulate as effectively as the fill-type insulation you buy. Common moss from ponds or lakes can be dried, possibly ground up in a feed grinder, and then packed into your walls to provide good insulation. When packing in pieces of moss or other materials that are not ground up, take care that no voids are left to create cold spots in the walls.

As a homesteader you most likely will shun credit and, conversely, some suppliers of credit will shun your efforts to obtain it. You then will have to live within your present means, and the type of buildings you elect to build will depend on your present financial condition. Most homesteaders will want to substitute labor for standard building materials as much as possible.

Basic Sheds

Like any homesteader, you will experience a certain pride of ownership in a sound, permanent structure built with little cash outlay. Poles or logs are the most basic and easiest-to-use building materials, if available. The simplest animal shelters can be made from poles lying horizontally to form three walls, with sod or straw around the structure for windproofing. Poles may be used to form a slightly sloping shed roof with branches, straw, and dirt on top to shed water.

For northern climates, enclose the shed with a fourth side and a doorway opening. Generally the doorway can remain open to a fenced run. For some animals, you may choose to hang a canvas or other wind protection over the doorway. Don't use a solid wood door, as your animals should have free choice of their shed or runway area.

Double-pole walls contain straw or other fill insulation. Another variation to the basic structure would be to trim the logs so they offer more resistance to air movement, as in log cabin construction. You may be able to find scraps of conventional building materials, like pieces of plywood. Don't hesitate to collect these when available.

Save wood by planting the poles of your structure in the ground. Use a windproofing and insulating material to fill the spaces between the posts supporting your roof. Wire fencing can hold a pack of straw, hay, or weeds.

If you have income or savings you intend to use in construction, consider buying polyethylene plastic 4 to 6 mils thick. Roll roofing or corrugated sheet metal are fine to use if you can afford them. Polyethylene film is about the best multipurpose material. It is vaporproof, windproof, rotproof, and affected by sunshine only over a period of time. When protected from damage by wind and sun, it lasts almost indefinitely.

Slab Siding

Another excellent building material available in many areas is slabs, the bark-covered outside portion of logs generally discarded when the interior part is cut up for lumber. Slabs make excellent siding for a pole structure insulated with straw.

Many mills will edge the better slabs at low cost so they are smooth on three sides and have bark only on the one curved side. These three-sided slabs can be used for almost any rough construction project with studs or light rafters. As siding on your house or buildings they have the attractive appearance of log construction. Unsided slabs often are used in roofs to hold up the layer of straw insulation.

Slabs make excellent fencing materials, as they offer wind protection for your animals. Remember that slab fences will catch drifting snow, so consider this in designing your pens. If you can get slabs at a reasonable price, you will have a good supply of a versatile building material and plenty of scraps and odd pieces for kindling your fires.

A hundred years ago, sod was the basic building material of grassland homesteaders. Where supplies are available, sod is probably most suitable for stacking around structures for its insulating value. Sod roofs are surprisingly waterproof when built up over shingles of birch bark that would otherwise blow away. Modern sod roofs are best built over polyethylene film.

Stone construction will usually be beyond the energy and talents of the average homesteader, although stone is one of the most permanent building materials. Stonework requires massive foundations, much time for collecting and preparing materials, a good deal of skill, and purchase of cement for the bonding mortar. It is unlikely that stone construction will be as inexpensive as other types, but you may have to give it a try if no other material is available.

Several magazines and catalogs advertise complete metal buildings ready to erect on your lot. These are suited for the suburban handyman, but are impractical for the homesteader. They are generally designed with the lightest materials possible to meet minimal snow and wind loads. Experience has shown that a high percentage of these small metal buildings become damaged from weather or use. The flimsy doors do not close properly after a while. The buildings require a better foundation than is usually provided, so this is an expense you may fail to consider when ordering.

Sounder structures for homesteaders can be made by purchasing standard building materials—but no decrease in strength or function occurs if your tool storage or other shed is made from rough lumber or slabs.

Building Codes

Despite what many people think, building codes were not adopted to harass people trying to build. Building permits and codes help you put up buildings that are safe and adequate for the intended purpose. They also help assure the person who might someday buy your homestead (or you, if you are the buyer) that the buildings will not fall down.

Generally, the codes require electrical wiring that will not cause fires, foundations adequate to support the structure, and buildings far enough apart for fire safety. The permits assure that a qualified representative of the granting authority has an opportunity to check over your plan.

Some states maintain authority for water, sewer, and other permits, while the county usually exercises building authority outside city limits. Almost all towns have building codes and require permits to build or remodel. If your land is outside the jurisdiction of any building authority, permits are not required. But once you determine that codes apply and permits are needed, be careful to follow the procedures outlined.

There is little chance you can get out of following the codes. Some rural residents may want to erect a unique structure, or use new materials or building techniques not accepted by their local code. If this is the case, you will need a variance from the chief building inspector or other authority. Be sure to obtain the required permits in advance.

Insurance

Whether you need fire and liability insurance is a personal matter. Some experts say every homeowner should have these basic types of financial protection. Certainly if you have a large investment in home and buildings, live near a town, and can afford it, insurance is a prudent purchase.

Consider buying insurance to protect your assets from two important threats: (1) the risk of *physical damage to your property* (i.e., buildings, contents, equipment, and revenue stream) by such causes as fire, lightning, windstorm, hail, flood, and other natural and man-made perils; and (2) *civil liability for injury or damage you cause to others*.

Property insurance (provided in a Homeowner's or Farm Package policy) can protect you against more than just fire. Think about your vulnerability to other causes of loss—for example, lightning, windstorm, water damage, equipment breakdown (including off-premises power outage), theft, and vandalism—and arrange coverage that addresses each of those risks. The broadest type of policy is a so-called "All Risk" property form, which insures against all causes of loss not explicitly excluded. [For example, the policy would not cover wear and tear, deterioration, lack of maintenance, etc.]

Liability insurance is virtually indispensible in today's litigious world. It's easy and inexpensive for someone to lodge a claim of negligence against you (for example, because of bodily injury or property damage that you supposedly cause), yet the legal fees to defend yourself against such an action can be significant—to say nothing of a potential verdict or court award. A liability policy covers defense costs, settlements, and judgments, making it a sound investment to protect your assets. There's no rule of thumb about how much coverage to buy; the answer depends on how much you have to lose in a worst-case scenario—coupled, of course, with the cost of the policy and relative value it represents.

Advice and Help

Your state university has research experts on agricultural and country living problems. At each university Cooperative Extension, specialists keep informed on research findings and the latest agricultural practices. They make information available to home agents and county Extension agents whose primary purpose is to educate the public.

Through the Cooperative Extension, your state university maintains a plan service, where you can purchase copies of all farm building plans considered appropriate for your state. Many U.S. Department of Agriculture and state bulletins are available there.

Banks and government lending agencies offer free counseling, especially if you apply for a loan. Libraries are prime sources of all kinds of information, where you can read copies of the latest how-to books and magazines.

Don't forget to talk to your neighbors. They probably know more about the local climate and other conditions than anyone else.

When you get close to actually building a new structure, sort through all these sources of information and settle on a final plan. Remember that each plan is a compromise of features and prices. Finalize, on paper or in your mind, just what you intend to build before you start. Occasionally you may find a plan just as you want it, but generally the plans of others can be improved upon for your purposes.

The plans and suggestions that follow may be just what you need, but more likely they will suggest ways to combine your materials and talents into the plan just right for you.

Building Barns

Barns are the classic rural American buildings for storing crops and sheltering animals. A barn need not be of the large size and classic shape most people visualize, but should be sized to meet your needs and financial situation. You will have to decide which farm functions you want confined to the barn and which should have a separate structure.

Use your barn for storage and animal pens. You can keep more than one kind of animal, if you have separate pens. Use outside pens and crude shelters for pigs and goats while keeping a few chickens and a cow in the barn. These outside animals can be moved into the barn during storms. You probably should have a separate feed storage room, even if it is small, to keep wild animals as well as your farm animals out of your feed.

Confine chickens in a good pen or separate room. It can be either wood or wire mesh. If predators like weasels or foxes are in the area, your chicken pen needs to be tight enough to keep them out at night. Use large windows in your chicken pen area or a plastic-covered wire mesh to admit sunlight. Space required for chickens can be as little as 2 or 3 square feet per bird, so that sixteen to twenty-five chickens fit in a 6-by-8-foot enclosure. Their crowded situation is helped by an outside run accessed through a small door.

If you plan to keep one cow, a second stall can serve for hay storage. Until you get your workshop built, reserve space for a work area out of the weather. Functions of your barn will change from time to time. As you move some activities to their own separate building, you will make space in your barn for other storage or activities.

Build your barn as large as you can afford, consistent with getting the construction done within your limitations of time and materials. It will be your largest, most expensive, and most time-consuming project,

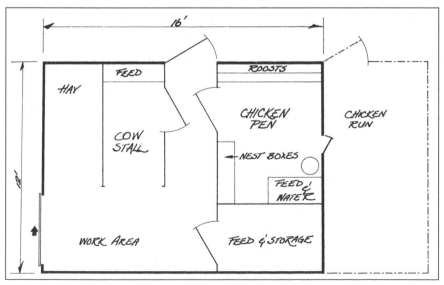

You can keep more than one kind of animal in your barn, as long as you have separate pens. Here, one cow can cohabit comfortably with a small flock of chickens.

after your home. Thus, you will have to consider whether you want to build your barn first or construct a pigpen, goat house, or tool storage at the outset.

Chicken Coop

Chickens need a well-lighted structure, as they are responsive to light conditions and day-length. You may provide artificial light in a basement or windowless insulated structure, or you can provide adequate window area for natural lighting. In northern areas your egg production will be better in winter if you provide a few hours of artificial light, but you can still maintain some production without it.

Chickens also have the advantage that in summer they can be turned out to "scratch" for themselves. If left to run in your garden, chickens may destroy small plants. Besides fencing a poultry run where you can throw a great variety of weeds and other food items, you may want to fence your chickens out of the garden. Thus they can have free run to forage your woodlot and fields without doing damage.

Small flocks usually thrive on a floor (which can be earth) covered with a litter of straw, rather than kept in cages as for commercial production. Your chicken coop will then have an area for roosts with a dropping pit beneath, and an area along one wall with built-in nests.

On your floor or suspended from the ceiling you need a feeder for chicken mash and a feeder or box for grit. A waterer or two completes the equipment. You can provide a "scratch" feed of cracked corn, wheat, or barley, simply by spreading it on the floor. The chickens will dig in the litter and help keep it stirred up and dry.

In the winter, it's not necessary to heat poultry houses for small flocks unless you live in a severe climate. Be careful if you use a space heater, as contents of poultry houses burn. If you keep the concentration of adult chickens at about $2^1/_2$ square feet per bird in a well-insulated house, the chickens will generate enough heat to keep the house above freezing in all but the worst weather. Build the small door to your chicken run so it can be closed.

You will be able to keep a few ducks or geese with your chickens if you choose. These will nest on the floor.

Housing for Pigs

Housing for your pigs can be crude, but provide them with a dry place to sleep and protection from drafts. In severe weather, pigs will burrow into their bedding until you can't see them at all. Due to their compact shape they can stand cold, but they can't bear prolonged hot spells without shade.

To get started you may want to purchase a weaner pig and raise it in a small shed in a fenced pasture or enclosed pigpen. Once you get used to raising your pig to the butchering weight of 200 pounds or more, you will know if you want to keep a breeding sow and raise your own litters.

To overwinter the sow, you will want a windproof shed. This is a chance to use imagination and ingenuity to construct a sound building from materials at hand. A low structure measuring 6 by 8 feet with a door 4 feet high and $2^1/_2$ feet wide suffices. For milder climates a tight shed that is uninsulated will do.

Goats, Pigeons, and Rabbits

Goats can use the same simple housing as pigs, or a larger barn shared with other animals. Build your goat shelter near a run or pasture with a high fence.

The so-called proper way to house pigeons is to build a loft with an attached enclosed flyway. But by the time you have finished the building, bought feeds and equipment, and paid for probably expensive breeding stock, you will decide that it will be a long time eating squab to get back your investment.

Alternatively, if your structure is such that the pigeons can be turned out to forage most of the year, you will have little feed cost and can show a profit more quickly. Pigeons have traditionally used barns or haylofts for shelter and soon learn if there are any feedlots, grain elevators, or other easy sources of feed nearby.

With this approach, you can adapt almost any shed or barn for pigeons. They require some nesting spaces and horizontal shelves for walking and resting. Since they are fliers, they will concentrate near the ceiling and need some roosts or landing areas. Pigeons tend to go back to the home where they were raised. Thus you will want to have a place to keep your breeders so they won't leave. Once the young learn where home is, you should have pigeons from then on.

Rabbits do nicely with minimum care in a shed or barn. If you want them confined, you will probably need a floor in your rabbit shed, as they like to dig burrows and will eventually get out. If they have straw or hay for bedding, you only have to put the buck in your rabbit pen occasionally, and keep them in feed and water, to have a batch of young rabbits about ready to eat at any time.

Rabbits need even less care when you turn them out in the spring, summer, and fall. They will provide themselves shelter under buildings and anywhere else they can hide. When fall comes and little green grass and garden remain, you can catch your breeding stock and confine them to the rabbit shed. Corner them by feeding them a little inside the shed.

Rabbits are prolific, so you will not have the expense of keeping many breeders over the winter. When your shed is warm and well bedded, you may even get a litter or two in winter to keep you in meat.

Housing Horses

If you have adequate acreage, facilities for putting up hay, and some pastureland, and want to keep a horse at minimum cost, then you will not need the complicated and expensive structures for keeping a horse on a small acreage. With free access to a 10-by-12-foot open-front shed that is well bedded, your horse will do fine.

A barn with a 12-by-12-foot space for a box stall or a 5-by-9-foot space for a tie stall is also adequate horse housing. Under these conditions be prepared to let your life revolve around the horse, as it will take most of your spare time and cash for its care.

Look for horse barn plans at your state university's Cooperative Extension offices. See also "Further Reading" (page 313).

Storage Sheds

Storage for tools, hay, machinery, and equipment needs only a roof, or roof and walls to prevent deterioration. Storage for tools and small equipment usually needs a windproof sidewall to prevent entry of dust, rain, or blowing snow. Prime needs are adequate shelves, hooks, pegs, or compartments to hold the stored items.

Store grains, dry feed in bags, and other food items in areas designed to keep out wild animals, mice, and rats. Wood buildings with a few cracks can be improved by nailing metal from tin cans over cracks and holes. Or else use 35-gallon garbage cans for feed containers, or construct mouseproof feed storage boxes.

Bear in mind the importance of keeping items dry, especially feed. If your soil is moist, build a floor in your storage area and allow air to circulate freely beneath the floor. This air circulation will carry away moisture from the ground before it can enter the storage area and cause problems. For more on storage sheds, see the chapter "Work and Storage Areas Outside" (page 64).

Water and Waste Disposal

To estimate water needs, consider the future. Whether you choose to install a new well, reconstruct an old one, tap a spring, or add water to your system from a cistern, reservoir, or storage tank, you should plan for any dream projects or goals as well as for your immediate needs.

Research studies have produced this formula for estimating home water needs: The largest water requirement for a single fixture (usually the bathtub or automatic washer) plus one-quarter the requirements for every other fixture (the kitchen sink, the shower, each toilet, etc.) equals your house's peak water demand—those periods when the well and pump must supply water continuously.

The unit of measurement for the formula will be gallons per unit of time, such as gallons per minute, or GPM. After you've established a reasonable GPM demand, a water source and delivery system can be set up with capacity to meet the demand.

How Much Water Do You Need?

Studies show that home water use has reached 80 to 100 gallons per person per day, an increase from the average of 50 gallons in the late 1970s. Many people use less, and some use more. Today's domestic water consumption is certainly more than necessary, because it includes such things as showers and frequent laundering in addition to basic needs like drinking, washing, and flushing the toilet. A bare minimum for people, including bathing and cooking, is between about 8 and 50 gallons per person per day.

You will think harder about this when you have your own well. Wells that meet demands of the farmstead home will usually also meet the water demands of a small-scale farm operation, one with just a few head of livestock, for example. For more elaborate projects, such as automatic stock watering or extensive irrigation, you will want to retool your water sources to meet the higher demands of the farm. (Worldwide, agriculture uses nearly 70 percent of the water people consume.) Or new water sources may have to be developed.

TABLE OF WATER REQUIREMENTS FOR INDIVIDUAL FIXTURES TO HELP ESTIMATE HOME WATER NEEDS	
Water Uses	Flow Rate in Gallons per Minute (GPM)
Household Uses	
Bathtub, or tub-and-shower combination	8.0
Shower only	4.0
Lavatory	2.0
Toilet–flush tank	3.0
Sink, kitchen–including garbage disposal	4.0
Dishwasher	2.0
Laundry sink	6.0
Clothes washer	8.0
Irrigation, Cleaning, and Miscellaneous	
Lawn irrigation (per sprinkler)	5.0*
Garden irrigation (per sprinkler)	5.0
Automobile washing	5.0
Tractor and equipment washing	5.0
Flushing driveways and walkways	10.0
Cleaning milking equipment and milk storage tank	8.0
Hose-cleaning barn floors, ramps, etc.	10.0

*Some irrigation sprinklers have more water capacity than shown in this table. If the capacity of your sprinkler is known, substitute that figure.

By having a private water system, you should know more about well construction and sanitation than city friends who depend on a municipal water system. Knowing your waterworks does not rule out the threat of waterborne diseases. If one factor in the system is most important, it is sanitary protection of the source of your water. Sewage, animal wastes, or chemicals of various kinds can pollute your water.

Newly constructed wells can lead to contamination of groundwater, unless you take precautions. In the process of drilling, boring, or digging a new well, natural earth barriers to surface and subsurface waters will be disturbed. The well itself can become a low-resistance path for contaminants to travel from ground level to below the water table. However, the path can be sealed off with a grout made of neat cement and water.

Your best assurance of the proper installation, materials, and location of the new well is to hire a licensed well driller. Authorized drillers know the water-bearing strata in their locality and should follow state health department regulations.

Sometimes groundwater can become contaminated from nearby wells that are old, poorly constructed, and unable to hold out surface drainage. Remember this model to help avoid contamination problems: Three things are needed for the entry of surface contaminants into the groundwater supply: First, a contaminant; second, a transmission path; and third, a transporting medium. For example, harmful bacteria (the contaminants) could reach your water supply after they travel through a well bore (the transmission path) in the movement of surface water or rain (the transporting media).

Rebuilding an Old Well

It may be more economical to reconstruct an old well than to install a new one. Before doing so, however, check to see if there are any obvious

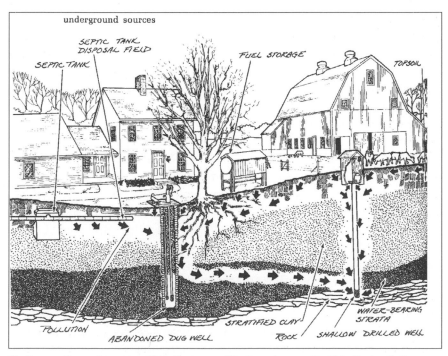

Shallow wells can become polluted more readily than deep wells. Note that pollution can come from underground sources as well as surface sources. (Adapted from *American Association for Vocational Instructional Materials*)

contamination sources near the well. Check the drainage. Will the proposed well structure be an adequate barrier against contaminants? Also before you begin, measure the yield of the old well. If the water is laden with sediments, clean out the well before measuring. You can normally have a rural property with an existing water source evaluated by the county health department.

The old well may have been designed for use before electricity was readily available for pumping. It may be inadequate after you add modern pumps and plumbing. But old wells often can be made deeper to reach a greater flow of water. With modern well casings and pumping equipment, a reconstructed old well no longer needs a large diameter. A new casing of 5 inches or less will meet farmstead water needs. In figuring the cost of renovating the well, or installing a new one, add the cost of closing the old well bore properly.

A typical procedure is to backfill the hole with sand, after placing the new casing and pouring the special well-grade gravel around intake screens at the base of the casing. As the sand is filled around the new casing, water will surge into cracks and crevices of the old walls, carrying sand with it. This improves sanitary protection of the new system.

Springwater

Springs, or natural flows of water from the ground, can also be developed as a water supply source. If you have a good spring on your property, ask these questions before acting: Is the springwater of good quality? Is the flow adequate? Could a gravity feed be set up from the spring? If not, would the cost of pumps and piping be within reason? And finally, because springs and areas of seeping groundwater are frequently flooded, can the water source be protected from contamination?

You should have control over any storage reservoir for springwater. If the spring discharges more water than you can use, set up a diverting device for water to go past the reservoir except when needed. Also, consider digging diversion ditches around any protective housing for the spring.

One more point on developing your water source: If you plan a business catering to the public such as a campground or a riding stable, your water system probably will have to comply with the federal Safe Drinking Water Act by following the latest regulations for well construction and sanitation.

Beyond your source of water, performance of the system is determined by other components, specifically the design, size, and maintenance of the pumps, valves, pipes, and storage facilities. A widely accepted term for storing water is *intermediate storage*. It applies only to drinkable water held at normal atmospheric pressure in a storage facility carefully designed and constructed to protect the quality of the water. A separate pump (besides the well pump) is needed to distribute water through the system.

With intermediate storage, a well yielding less than 1 gallon per minute can provide good to excellent water service for a home. A 3- to 5-GPM well can provide enough water for about an acre of lawn and garden besides home usage. Intermediate storage at or below ground level usually provides the best per-dollar storage. The water is protected from frost, and in summer comes to the spigots at cool temperatures. On the other hand, water stored in overhead tanks or held under pressure can be over ten times more expensive. The water tower or stanchions are costly. And with pressurized tanks, not much water can be held at reasonable cost.

Firefighting

Stored water can serve the farmstead in many different ways. It is available during power failures, or used for fighting fires, or for first aid in emergencies. The local fire chief can help you plan adequate storage to handle fires on the farm. For example, if you have 5,000 gallons available, a rural fire department pumper can fight with 500 gallons a minute for ten minutes.

Lack of available water is a major problem in rural fire control. Sometimes even if water is at the scene, it can't be delivered when friction in the plumbing reduces the water pressure. Friction loss actually affects more than fire protection. Even with adequate well-pump capacity and intermediate storage, friction loss from poorly designed plumbing can mean the difference between a high-performance water system and an inefficient one.

In the design and construction of private water systems, too much emphasis is placed on quantity and quality of the water and not enough on quality of the service. It is unpleasant, to say the least, when competition for water between fixtures becomes a major part of planning household activities.

How would you change your water system if given the chance? Most farmers who cooperated in a dairy farmstead study in the 1970s answered by saying they would like to be able to take showers and be totally unaware of water use in the kitchen.

Replacing Valves

Sometimes much of the friction loss in older water systems is caused by undersized globe valves, also called compression stop valves. They

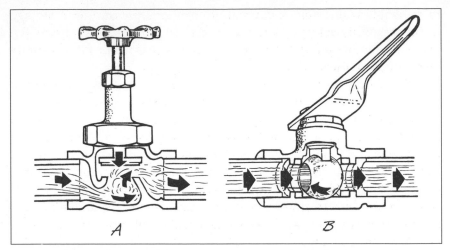

Sometimes pressure loss in older water systems is caused by the friction created in undersized globe valves (left). They should be replaced (right) with ball-type valves. (Adapted from *American Association for Vocational Instructional Materials*)

should be replaced with ball-type valves. Small-diameter pipe can also contribute to friction loss. For underground pipes, use at least 1¼-inch pipe. You won't save cash with smaller pipe—the same size trenches must be dug.

Water for farm work is related to labor costs. Even with no hired hands, when you need a certain amount of water, you want it pronto! You don't want to waste time and money waiting for it. Therefore, it is wise to minimize friction loss.

With good planning and management, you can irrigate a large area with sprinkler technology at a reasonable cost. (Energy requirements will normally be less than 1 kilowatt hour per 1,000 gallons of water for a well-designed system.)

If you choose crops that have critical watering periods at different times, a larger area can be irrigated than your water supply might indicate. A small deep well pump, intermediate storage, and a small pump between storage and sprinklers can do the job.

Water from surface sources such as streams, ponds, lakes, or rainwater stored in cisterns should only be used if groundwater is unavailable. Any surface water will be polluted. Contact your local health official about proper treatment and disinfection. Tapping surface waters may also involve special water rights problems, so investigate before proceeding.

Creating a farm pond is another way of getting water, especially for nondrinking purposes. You should have full control of the water coming into the pond. If the pond is to be fed from a stream or spring, this may mean excavating for the pond above the level of the source and pumping water up to it. If a watershed drains into your pond, keep the watershed grassed, and free of barns and septic systems. The pond spillway—that part of the pond's banks or dam where excess water exists—should be large enough to handle flooding from heavy rains.

Waste Disposal

As a rural resident, your responsibilities broaden. Besides meeting the needs of your family and the immediate community, you are now a steward of the land and the community in a larger sense. Most pertinent here is that the farm be managed so that you do not waste water and that you don't pollute it with an ill-conceived waste disposal system.

The nineteenth-century invention of the septic tank–soil disposal system brought the indoor toilet to rural America. A properly functioning, properly located septic system is still a very efficient way of disposing of sewage. Today, it remains the most common type of private sewage system.

The septic system consists of two parts: the septic tank and the solid disposal area, also called the drainage or distribution field. The septic tank is where sewage solids separate from liquids and where bacteria begin to decompose the solid material. The soil disposal area performs most of the sewage treatment. Starting at the outlet pipe of the septic tank, a system, usually consisting of 4-inch corrugated plastic pipe, extends through trenches containing gravel buried beneath the ground's surface.

Discharge from the tank is a gray, somewhat odorous liquid, referred to as septic tank effluent, which carries suspended solids. The effluent goes through the drainage system and passes through holes in the piping. Soil bacteria and fungi then decompose the solids into inert matter. The result should be a clear, bacteria-free, and odor-free effluent. But stop—that is only how the system *should* work.

Although the septic system is an old idea, many systems have been poorly designed and constructed. A large portion of those in operation today simply do not work properly, and are serious threats to health and environmental quality.

Clogging Is the Villain

Research indicates that with proper design and management, modern septic systems can have a nearly infinite life. The research shows that most premature failures are due to clogging of the soil disposal area. Breakdowns can also result from poor septic tank performance, high soil moisture during construction of the system, failure to properly evaluate

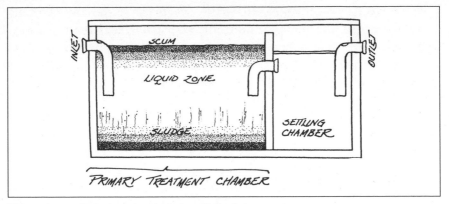

The anatomy of a two-compartment septic tank.

soil drainage, or overloading the system. In short, septic system failures are caused by lack of foresight or by neglect.

The inlet to the septic tank should be nonfouling and designed to cause a minimum of turbulence in the tank as sewage enters. From the intake, heavy sewage particles settle into a layer of sludge at the bottom of the tank, and lighter-than-water substances, such as grease and fat, float to the top, forming a layer of scum.

Between the scum and sludge layers, there is a zone of clear liquid. The outlet, located at the upper level of the liquid zone, should permit only a minimum of solids to exit onto the soil disposal area. However, especially with single-compartment septic tanks, turbulence from incoming sewage, gas bubbles rising from the sludge, and the like may allow too much suspended solid material to leave the tank. The soil disposal area then becomes clogged.

Many tanks have two compartments, the second serving as a settling chamber. These tanks can reduce the amount of suspended solids discharged into the soil disposal area by 25 to 30 percent. For any given design, the larger the tank volume, the more suspended solids will be removed.

Until you are familiar with how often your septic tank needs pumping out, have the system inspected once a year. Part of the scum and sludge cannot be eaten by bacteria and must eventually be cleaned out.

Because the septic action of the tank depends on bacteria, nothing should be flushed into the tank which will kill or retard growth of the bacteria. They will tolerate moderate amounts of soaps, detergents, disinfectants, and similar household products. But don't overdo it.

The septic tank should be located on the downhill side of the water source and at least 50 feet away. Size and location of the soil disposal area depends on the terrain and capacity of the soil to absorb sewage liquid, the effluent. Heavy impervious soil, a high groundwater level, or insufficient land area can create problems for the soil disposal area. When soil around the disposal area becomes saturated with moisture, for any reason, decomposition of solid particles in the effluent slows down. The system clogs. Effluent can back up in the septic tank. Groundwater can be harmed.

The problem may seem a physical one, but it is really a matter of the wrong chemistry. Organic compounds in sewage or septic tank effluent need lots of oxygen to decompose properly in the soil. When the soil is loaded with moisture, not enough oxygen is present and the system stalls. New ways have been developed to keep the purifying action going under difficult circumstances, and still avoid groundwater spoilage.

Pressurizing

One way is to pressurize the system. A time-controlled pump sends effluent into the disposal system for only brief periods at a time. Corrugated drainage pipe is not used. Instead, the pump sends effluent through 1- to $1^1/_2$-inch-diameter plastic pipe, which has small holes every 30 to 42 inches. For sandy soils, the pump should be set to eject effluent into the pipes four or five times a day. For heavier soils, just once a day may be enough to prevent clogging.

Another way is to alternate between two or more drainage fields by using diverter valves. Alternating offers the advantage of giving the soil of each disposal system a rest. If one system clogs, it can immediately be shut off without stopping operation of the entire septic system. For many soil types, one disposal system can be used for six months to over a year, before switching to a fresh system—one with enough oxygen in the soil to properly decompose sewage.

Serial Distribution

A third way, the serial distribution system, has proved very effective on soils where there is no risk of polluting the groundwater. A series of trench sections are separated by dams or earthen barriers. As effluent flows into the system, the first trench section is forced to fill completely before the liquid can go to the next section.

Flooded sections serve as contact chambers, where bacteria decompose suspended solids. The biological activity continues to improve effluent quality in each succeeding section. This system is often used on hillsides, but it is equally effective on flat land. If the serial system should fail, only the first trench usually needs to be replaced. The series of trenches can also be alternated.

The pressure distribution system could be used for subsurface irrigation of shrubs, trees, and lawn. Normal amounts of effluent entering the soil from a septic system far exceed what landscape plants would require.

If you are hooked into an expensive town water system, consider making the soil distribution system larger, with separate laterals and two-, three-, or four-way diverter valves to accommodate the irrigation. Plant nutrients in sewage effluent make such a system worth considering.

Recycling and Composting

No discussion of waste disposal on the farm would be complete without touching on garbage disposal. Two things here—the compost pile and waste recycling.

Recycling has become a way of life in most regions of the United States. In your country shed, you probably will need separate disposal containers for paper, glass, metals such as aluminum and steel, and plastic bottles.

Composting is an ancient art, and a modern science. On the farmstead it can be used to provide fertilizer for the garden or it may be an integral part of raising crops, depending on how much organic material is obtainable.

Layer leaves, hay, sawdust, crop residues, weeds, grass clippings, kitchen wastes, manure, and other organic materials in piles. Over a period of several weeks, they ferment. With the aid of oxygen, provided by turning the piles occasionally, the material breaks down quickly into an excellent soil conditioner or fertilizer.

A small compost pile can keep a vegetable patch in top condition. If a lot of composting material is shredded, piled, and turned with power equipment, orchards and field crops can also benefit from the additional humus and fertility that compost gives to the soil.

Power Sources and Equipment

The small country site opens new energy sources to the owner. Natural energy from wind, water, the sun, wood, or coal may be locally utilized. Electricity, natural gas, fuel oil, and coal may be supplied by commercial companies.

Equipment such as tractors and tractor-operated implements, household appliances, and power tools use many sources of energy. Families living alone require surprisingly large quantities of energy. More than 20 percent of all the energy in the United States is consumed in the home. Over half of this energy is used for heating.

Both your lifestyle and the nature of any operations on your acreage will determine energy requirements. The location of your place will determine the most likely energy sources. Most small farms will be near electricity or a petroleum fuel source. However, the electrical service may be inadequate for large electrical motors, particularly three-phase motors. Accessibility for fuel deliveries may be poor during some months, and larger storage facilities may be needed.

Electricity obtained from a central station is fairly dependable and reasonably priced as a rule. Electrical generators on your few acres are ideal for standby operation. If generators are your only source of electricity, some form of energy storage is needed; direct current, or d.c., and storage batteries are used. These systems are satisfactory for electric lights but have limited capacity for large power demands.

Other sources of commercial energy that may be available at your site are natural or LP gas, gasoline, fuel oil, or coal. All of these fuels can be used for both power and heat. Gasoline and diesel fuel are used regularly for tractors. Any vehicle that runs on diesel can run on biodiesel. Some farmers make their own biodiesel with used vegetable oil from local restaurants, but many areas now deliver biodiesel. LP gas is normally used for heating, but can be used in cars, trucks, and tractors fitted with special LP gas carburetors.

In select locations, water and wind energy may be useful for performing some operations. Work on utilization of wind energy has come from many sectors and may result in equipment directly coupled to a wind turbine, or in wind-generated electricity.

In 1850, just 1 percent of the total energy consumed in the United States was supplied by wind. Today, wind power on a larger scale is becoming more popular—and controversial in some areas. Even so, it still provides just over 1 percent of this country's energy.

The amount of power available is directly proportional to the cube of the wind velocity, which illustrates the importance of wind velocity. Doubling the wind velocity means 2 x 2 x 2, or eight times the power. As a rule of thumb, continuous winds at an average of more than 8 miles an hour are needed to operate a wind-powered electrical generator.

Suitable site characteristics include high annual wind speed and no tall obstructions upwind for some distance, depending upon the height of the wind turbine. They do well on top of a smooth well-rounded hill, open plain, or shoreline, or near a mountain gap that produces a wind funneling. Consult the turbine manufacturers for specific information.

Waterpower is another potential source of natural energy available at some farm sites. In the past, waterpower provided the energy for grinding flour, sawing wood, and generating electricity. Today waterpower is used primarily for generating electricity in large central hydroelectric plants. Very few of the original small-scale systems remain. Interest has developed in harnessing water in rivers and streams for limited supplemental energy.

Wood and Solar Energy

Wood is a source of energy available at many sites. While only a small fraction of America's fuel needs are now supplied by wood, it can be used advantageously if a sufficient, low-cost supply is available. Few fireplaces built to look pretty can supply real heat, as the efficiency is a low 10 percent. Wood (or pellet) stoves warm efficiently. Use newer stoves that meet the U.S. Environmental Protection Agency's standards for after-burners, which combust more of the wood, reducing particulate pollution. Get wood from your own well-maintained woodlot or from forestland nearby. State and local regulations will govern wood collection from public lands, where a permit may be needed. Pellet stoves require commercially produced pellets. Outdoor wood furnaces heat efficiently and move the smoke and ashes away from living space, helping asthma and respiratory disease sufferers. But many outdoor furnaces are large, and all have low chimneys that make it necessary to locate such furnaces far from neighbors. If oil prices stay high, the furnace can pay for itself in about four years.

Solar energy has been used in recent years to heat and cool buildings, dry agricultural products, power irrigation pumps, generate electricity, heat water, and for other purposes. Solar cells are used to generate electricity from solar energy. Costs of solar cells and associated equipment are expected to come down with time.

Passive solar systems make use of the building design to capture and store heat. Windows or structural components that absorb heat energy are examples. Passive systems are relatively inexpensive and use few or no moving parts. They are designed into the structure and not added onto it later.

Active solar systems use large heat-collecting panels, pumps, or fans along with storage facilities. The equipment is relatively expensive, requires space, and must be maintained. While the technology has improved since the 1970s, when President Carter first installed solar panels on the White House roof, it is still not part of the mainstream, and it can't provide all of a house's heat. You will need a backup conventional heating system to provide 50 to 70 percent.

You can build a system yourself, but you must be capable of doing the job. Two rules of thumb on sizing the solar heating system are: first, that the collector area is equal to one-fourth to one-third of the house floor area, depending on the site and geographic location, and, second, that a liquid heat storage system needs 1.5 to 2 gallons of fluid for each square foot of collector, while an air heat storage system needs $\frac{1}{2}$ to $\frac{3}{4}$ cubic foot of rock per square foot of collector.

The house should be insulated to standards equal to electric heat. Often the many possible conservation practices already in place provide sufficient savings in an existing house so that the additional expense of putting in a solar system does not give a good economic return. The fastest payoff comes in heating household water. No structural insulation is needed, and comparatively small collectors—30 to 100 square feet—are used. The storage tank is connected to the household hot-water heater by a heat exchanger. Fluid in the collector does not mix with the household water, so there is no contamination.

Field Machinery

A wide range of tractors and implements are available from local farm machinery dealers. Most of the larger tractors and implements sold are manufactured in the United States, while many of the smaller tractors—particularly diesels—are imported. Small compact tractors, frequently called lawn or garden tractors, are also manufactured in the U.S. Farm machinery dealers or other specialized stores sell them. Garden tractors are designed primarily for light, estate duty and are not intended for continuous heavy service. Consider product reliability, equipment warranty, dealer reputation, and availability of spare parts when you purchase machinery.

A successful farm business requires careful management of land, labor, and capital. In most cases, machinery costs are high and exceeded only by land costs; thus, it is important to manage machinery properly. This includes planning the use of machinery for timely and productive operation, selecting proper types and sizes, proper care to maintain performance and reliability, replacing obsolete or worn-out machinery, and deciding whether you should hire a custom operator or lease a machine.

Ideally, each crop should have its own set of specialized implements to produce maximum yields. More equipment in turn means higher overhead costs. The lack of adequate equipment can delay getting crops planted or harvested on time, reducing yields and product quality.

Using larger machines reduces labor costs since they complete the job faster. But while large tractors can cover more acreage than smaller ones, they also have higher overhead costs. Smaller tractors have less capacity and may cause delays in key field operations, resulting in a lower crop yield.

Analyze the work you must do and the time you have available so that you can select the right size equipment. Farm tractors and equipment are commonly measured in terms of horsepower or machinery width. Machines should be selected on the basis of their capability to perform a given job within a certain amount of time. This rate of performance is called field capacity, and is measured in either acres per hour or tons per hour. Some harvesting and processing machines are measured in bushels per hour, but this is an inaccurate measure due to variations in grain moisture and densities.

Field capacity is based on actual time spent in the field. If it requires five days, working eight hours a day, to plow an 80-acre field, the field capacity is 2 acres per hour. Field capacity equals 80 acres divided by forty hours, or 2 acres per hour.

Field efficiency is the ratio of the effective capacity of a machine to the theoretical capacity. It measures the relative productivity of a machine under field conditions. Operating the machine at less than full width, lost time spent filling hoppers, cleaning, and adjusting machines in the field—all reduce field efficiency.

Some of the time lost in doing field work cannot be eliminated. Other lost time can be substantially reduced by careful planning and good management.

Servicing Machines

Keeping farm machinery in top mechanical condition is one of the best ways to improve field efficiency. Machines should be serviced regularly and adjusted correctly. Neglecting this can cause expensive repairs or complete overhauls. An operator's manual provided with all new machines gives proper instructions for servicing, repairing, and adjusting. Manuals for older equipment can usually be obtained from the manufacturers. Taking time to read the manual before using the equipment will eliminate future problems and save you money.

Replace farm machines when they become unreliable for completing the job on time. It is difficult to keep a worn-out machine repaired once it starts having excessive breakdowns. When purchasing a replacement, consider a used machine. Frequently machines in good mechanical condition have been traded in because larger equipment was needed. If these fit your needs, they can be a good buy.

Leasing machinery or hiring custom operators are alternatives to owning farm equipment. In some cases custom operators can complete the work faster and cheaper than you can. This is especially true when you have only a few acres and specialized machines are needed. When considering hiring a custom operator, talk to other people who have used his services. Waiting for a custom operator to arrive can be expensive if the crops are not planted or harvested at the optimum time. Timeliness is important when you compare leasing equipment, owning equipment, or hiring a custom operator.

Comparing Costs

Determine the costs of owning and operating machines before comparing viable alternatives. Machinery costs involve both fixed and operating costs. Fixed or overhead costs include machinery depreciation, insurance, interest, shelter, and taxes. Operating costs include fuel, labor, lubrication, maintenance, and repairs. Fixed costs generally depend upon how long a machine is owned while operating costs relate to its use.

Depreciation is a loss in the value of a machine caused by age, wear, and obsolescence—often the largest machinery cost. There are several different ways to calculate depreciation, including Internal Revenue Service methods.

Premiums paid for machinery insurance are included in the total costs of owning farm machines. Frequently, only the more expensive equipment is insured while a calculated risk is taken on less expensive machines. Typical insurance costs run about 0.3 percent of the original list price. Interest on borrowed capital or cash used to purchase farm equipment is a direct expense. Prevailing interest rates are used to calculate this machinery cost.

Storing machinery under shelter is a good management practice in most of the U.S. Studies on expensive tractors, combines, and other machinery have generally shown that storing these machines adds one or two years to their useful life. Typical shelter costs are less than 1 percent of the machine's original cost.

Taxes paid on machines are included in the cost of owning them. A typical annual charge for taxes is 1 to 2 percent of the machine's value at the beginning of the year. Typical fuel and lubricant needs amount to between 20 and 30 percent of the total machine cost. The ratio of fuel and lubricant cost to total machinery costs increases as annual use of the machine increases.

Fuel use can be reduced by matching tractor horsepower to machinery size. Other fuel-saving tips include eliminating unnecessary tillage operations, combining operations so the tractor can operate at full load, and throttling back and shifting to a higher gear when pulling a light load.

Labor to operate farm machinery is an operating cost and should be included in total machinery cost estimates. Either hourly wages paid for hired labor or a fair value for your own time should be used to determine the cost. This cost is particularly important when comparing custom machinery costs with owning or leasing machinery.

Maintenance and repairs are considered operating costs since the amount spent for repairs is proportional to machine use. While some repair costs are caused by deterioration, the majority are due to routine use of the machine. These expenses are almost impossible to predict because of the many factors involved, although high repair costs can almost always be traced to excessive speed, overloading, poor daily maintenance, abuse of equipment, and ignoring the first signs of problems. Repairs should be made to maintain the farm machine's reliability and keep it performing at top capacity.

Mowers and Lawn Tractors

Small equipment will meet the needs of some activities on an acre. Large lawns and small gardens require powered equipment, but not the large equipment used for field operations. Riding mowers will handle the grass-cutting chores. A garden tractor offers more versatility to do mowing, snowblowing, plowing, rototilling, and many other tasks. The bigger the job and the more tasks you have to do, the greater the horsepower and special features the tractor will need. Lawn tractors are light duty, while a garden tractor is often heavier duty and more capable of handling attachments that plow, disk, or till the soil. Some manufacturers distinguish between the two types.

These tractors are commonly powered by air-cooled single-cylinder gasoline engines ranging from 7 to 14 horsepower. Above 14 horsepower, the machine will use a two-cylinder air-cooled or four-cylinder water-cooled engine.

Electric-powered compact tractors are available and operate with minimum noise and vibration. The batteries are recharged by plugging into a 115-volt convenience outlet. Electric-powered tractors do have some drawbacks, including relatively short operating time between battery recharges and generally greater expense.

While manufacturers have made some efforts to standardize attachments, the best bet is to purchase the same brand of tractor and attachments. Some attachments are not available for the complete size range of compact tractors, since the smaller tractors do not have enough horsepower to operate them satisfactorily.

Small 20- to 45-horsepower agricultural tractors are available from local farm machinery dealers. These tractors are specifically designed for continuous-duty field work and cost more to buy than compact garden tractors. They should be equipped with a three-point standardized Category I hitch for attaching the various implements.

Three-point hitches provide near-parallel lift of the attachments, give better depth control of tillage implements, and provide weight transfer for heavy plowing. Standardizing the hitch lets you attach any brand of Category I implement to any Category I tractor.

Options available usually include a choice of either diesel or gasoline engines. Diesel engines cost more to buy initially but require less maintenance and use a lower-priced fuel. Consider the total number of hours you plan to run the tractor each year before deciding on the economics of a diesel versus gasoline tractor.

Bear in mind that many of these small agricultural tractors—particularly the diesels—are imported. Before purchasing, make sure your dealer has adequate repair parts or that they are available from some central location in the U.S. Spare parts shipped from abroad take longer and can delay getting your tractor repaired and back into use.

Small Tools

Buy a wide selection of hand and power tools at local garden centers, hardware stores, and mail-order houses. Select good-quality tools, as they will give better service and with proper care will last longer than less expensive equipment. Name-brand tools are usually more reliable, but you should shop comparatively, noting the best features and quality of each.

Machinery is potentially dangerous. Treat it with respect. Farm machinery designers have built in many safety features, but it is impossible to protect an operator who disregards all safety precautions. Practicing good judgment and common sense is the best protection.

Safety precautions you should follow include:

- Keep equipment in good operating condition.
- Don't operate machinery on steep-sided hills or near ravines.
- Use a tractor large enough to handle the load—a tractor can pull a heavier load up a hill than its brakes can control when coming down.
- Don't turn at high speeds—centrifugal force can cause side-tipping.
- Don't rush to finish a job and become careless.

Machines are built to help you. They will do an excellent job if you handle them carefully and treat them properly.

PART 4

The Big Picture

Five Years on Five Acres

By Arthur T. Johnson

Editor's Note: Arthur T. Johnson is a professor of biological resources engineering at the University of Maryland. In 1978, when the original Living on a Few Acres *was published, he had a 5-acre farm with everything from cattle to strawberries. Today, he specializes in fruit and sheep on 42 acres. Below is his 1978 essay full of advice based on his early years. We have updated any cost estimates to reflect today's prices.*

The first five years I lived on a small farm, I raised beef, sheep, pigs, chickens, ducks, geese, turkeys, vegetables, strawberries, and fruit trees. I tried various ideas to find the best, and I also read and compared the experiences of others.

Although living and producing on 5 acres is not the same as on 500, many techniques and methods can be adapted for small-scale operations. Keep in mind some basic guidelines when starting to farm on a few acres.

The first is, be cautious. Begin by producing on a very small scale, perhaps only enough to satisfy your own needs. Learn what problems you will encounter: diseases, insects, climate, and the requirements of quality and timeliness. These problems are likely to become more severe as production scale increases.

An important factor is whether you can produce your product profitably with acceptable quality, and what the local market is for that product. Some simple record keeping will be necessary, or you may end up subsidizing the food budgets of your customers.

Pleasing the Clientele

Pleasing your customers so they will return requires you to sell the best product you can. One man was loading a pig shoulder he had just sold onto his truck, when he accidentally broke it. So he replaced the pork shoulder with one from his own pork to satisfy the customer.

Knowing your market is probably the most important factor in developing a profitable enterprise. Determining where to sell your produce takes some work, but it is very important. Markets may be already available (a local farmers' market, for example), or you may have to create your own. This requires some selling on your part to get started, such as advertising in a local paper, and some system to keep in touch with past customers. Satisfied customers make the best advertisement for you.

There are economies of size. Small-scale farms cannot compete against large farms if they sell to the same markets. Market directly to the consumer, charging a fair price above wholesale and below retail. With food items, there is a considerable spread between farm market value and store shelf price. You and the customer can both do well by a direct transaction.

Selling directly to the consumer requires you to produce consumer-oriented crops, not field crops such as wheat. Strawberries, vegetables, fruits, honey, eggs, and popular meat animals are among the most marketable items. Cater to needs of the local populace, and grow items that are not abundantly available in your local area.

A good source of information on all matters is your local Cooperative Extension agent. Many of them are becoming more aware of the needs of small-scale farmers. Your state department of agriculture may also help.

Production methods may vary considerably depending on the crop and farm size. With your main source of income not derived from the farm, you can experiment with new ideas. Mistakes may not be quite as costly as on large farms.

Regardless of your small-scale farm's size, a considerable investment is still required. For example, rearing ten feeder pigs to slaughter costs at least $540, counting feed, the pig, depreciation of buildings, and other costs. An investment for dwarf fruit trees might be $30 per tree and 5 years to production. In general, the more profitable the crop, the larger the initial investment required.

Techniques

Decide on appropriate production methods. You can work on a large scale—such as buying a tractor, mower, rake, and baler to produce hay—or small scale—such as using the lawn mower to chop grass for silage. To control weeds in a strawberry patch, you might compare the merits of a large tractor and cultivators with a rotary tiller; with simply herbicides; or finally, with the most simple alternative—mulch.

I prefer a balanced farm with little waste. This means several crops, both animal and vegetable. Pig waste can be used to help grow strawberries and vegetables; waste vegetables can be used to help feed pigs. (If you use pig waste, apply it in the fall.) Ground unsuitable to vegetable growing can be used as pasture for sheep, goats, or cows, and fruit trees can grow in the same area. Ducks, geese, and chickens help fertilize the area. You can sell all of these.

In all of this, you must have tools and equipment. To reduce your investment, watch for announcements of farm auctions and sales. Used tools cost a fraction of the price of new ones.

Remember that laws and regulations govern the production and sale of food, and that these vary from location to location. Often the sale of processed food is prohibited unless the processing was done in inspected facilities. Examples are butchering and cutting of meat and poultry, and handling of milk.

You may use pesticides, herbicides, other sprays, and additives within registered restrictions (crops, application methods, withholding periods). Beehive inspections are required. Use of sewage sludge for fertilizer may be restricted, while use of composted sludge may not be. Know the regulations that apply in your area, especially if you deal directly with the public.

Small-scale farming can be a rewarding hobby. Often it requires most of your free time and does not allow you to leave home for extended periods—but for those who like it, every day is a vacation.

Welcome to
SweetAire Farm

By Arthur T. Johnson

Editor's Note: In this new essay, Johnson reflects on the last quarter century: his decisions to add acreage, concentrate on fruit, and keep his day job.

I am a farmer part time. That means I have two full-time jobs, the farm being one of them. My farm is called SweetAire Farm and is located just outside the little town of Darlington on the edge of the rural-suburban interface in Maryland. That means that where the farm is located is rural right now, but there are signs that urban encroachment is coming fast from two directions: from Baltimore to the south and from Philadelphia and Wilmington to the northeast. In a few years there may be farms scattered among suburban developments as there are in many locations on the East Coast.

This patchwork pattern isn't all bad. The proximity of the population provides a ready market for direct marketing of farm produce, and that's good. Close neighbors don't always understand or tolerate farming operations, so that's bad.

My family starts with my wife, Cathy, and four children: Joy, Jodi, Paul, and Eric, in that order. The children came in two bunches of two each. Joy and Jodi were in their mid- to late teens when we adopted Paul and Eric as youngsters. This means that the flavor of SweetAire Farm has changed over the years, as children with different interests combined with the enlargement of our property holdings, and as we learned from our foibles. This will be made clearer in the pages ahead.

SweetAire Farm is a general family farm of 42 rolling acres. That's not really big for the area, but it's a lot larger than many others. On this farm we raise some cows, sheep, hay, and fruit—mostly fruit. Fruit provides us with most of our income, as well as most of our opportunity to work.

We have (as of this writing) about 1,400 fruit trees of all kinds: peaches, apples, plums, pears, nectarines, and cherries, in order of decreasing numbers of trees. We have many different varieties of each, including some that aren't normally available commercially, and including many that people don't normally see in food stores. In addition, we have strawberries, grapes, kiwis, rhubarb, blackberries, blueberries, currants, and other minor fruits. We sell these at farmers' market and sometimes at home. We sell strictly retail; wholesale prices are too low to make these sales worthwhile.

We have about ten Suffolk sheep. Ten is all we have adequate pasture for, and ten is all we have time for. Sheep demand a lot of time and effort at certain times of the year, so we don't need too many.

We have six to eight Hereford beef cows, with mostly reddish brown bodies and white faces. Again, any more than this would require a lot more pasture and hay than we are prepared to dedicate. We put up about 800 bales of hay a year to supply the cows and sheep. That is not a lot of hay for a farm, but haying requires a lot of intense effort and doesn't result in a lot of income, so I'm not anxious to increase this amount.

In order to help run this farm, we have four tractors, a truck, a mower-conditioner to cut and crimp hay, a hay crimper, a hay rake, a baler, wagons and trailers (the wagons have four wheels and the trailers have two), two tractor-mounted rotary mowers (5 and 10 feet wide), a cherry-picker lift, a sprayer, and other machinery. We have two rather small barns, a machinery shed, other storage sheds, and several other outbuildings. My motto is, "When I run out of storage space, I build another building."

SweetAire Farm started out as a 5-acre piece of property with a house, one small barn, some sheds, and space overgrown with briars and small trees. We cleared the brush and trees, built some fences, and made some pastures. Then I began to plant fruit trees, grapevines, and other fruiting plants until I was tucking something into every available space. We were bursting out of our seams.

By this time we also had sheep and cows, with no place to grow hay and with inadequate pasture. In 1979 we bought 15 acres located $^1/_2$ mile down the road from the house. That property had been used to grow corn, and the ground was still bumpy from where the furrows had been plowed and not quite smoothed. When we bought the ground, however, it had been neglected for several years and was even more overgrown than was our original piece of property.

We had big plans for that 15 acres: We could use some for pasture, some for hay, and we even had visions of pick-your-own strawberries. But first, the trees and brush had to be cleared.

We had very little in the way of machinery at that time, so we cleared it by hand. The small trees were either clipped with shears or cut with a saw. The brush was also clipped to the ground. All of this brush had to be moved by hand to the hedgerows. The girls helped a lot with that chore.

One of the best ways to keep the trees and brush from regrowing was to cut the cleared area for hay. We did this with an old John Deere "A" tractor and a sickle-bar mower, bought very used from a farm machinery dealer. We raked by hand and pitchforked the loose hay onto the truck or, later, the trailer.

As time went on, we made progress. Things we had done by hand we did by machine. Things we did with simple machines we did later with more sophisticated ones. Once cleared, the brush did not grow back. Once built, fences remained. This was bootstrap farming at its purest. We bought what we could afford, and only when we needed it. Except for a few occasions when we needed larger tractors, we did not borrow money to finance our purchases. We didn't know if there would ever be a financial return on our investments, although that was our goal. At the time, we did it because we liked farming and wanted our children to know what farming was about.

We were feeling our way. What methods worked best for us? What things could we do that would be successful? What could we sell that other people might want to buy? What could be possible with the limited amount of time and money that we had to work with? How could we realize some return?

We had many choices. Farms in the Maryland countryside are lush and green. The soil is fertile and rain is plentiful. Maryland winters are cold, but not too cold. We normally have several weeks of snow that begins accumulating after the first of January, but the snow usually melts by the end of February, when the crocuses bloom. Maryland summers are hot and humid, but oppressive weather is usually confined to parts of July, August, and sometimes June and September. Springs and autumns are several months long and would be nearly perfect except for the excessive pollen that sometimes fills the air and overwhelms the sinuses. Rain is usually well distributed throughout the year, but short periods of dryness sometimes occur. The sun shines a lot.

Maryland is at the southern edge of the growth zones for northern trees and crops, and at the northern edge of the growth zone for southern trees and crops. All this adds up to a climate that stimulates grasses, bushes, and trees to grow with vigor. The proximity of the Chesapeake Bay;

Susquehanna, Patapsco, and Potomac Rivers—and numerous streams and ponds—even makes aquaculture a good possibility. So, the location and climate do not limit the choices of crops. Almost anything would be possible, if we could only find something that would work for us.

We tried raising cows, sheep, and pigs for sale, and had some success. We had regular customers and a stable market, but the financial returns were modest, while the amount of work was relatively large. Besides, there was always the necessity to sell to the customers, and that required a kind of effort that I did not find easy to give. So, we tried other things.

We considered selling hay. If the hay quality is high enough, there is a good market to those who own horses for pleasure. Those people want to ride and pet their horses, but they don't like to clean out their stalls. They absolutely do not consider putting up hay.

But hay requires manual labor, and hay must be stored. The best time to sell hay to minimize storage space is in the summer, when it is in the field. However, summer is when the pasture grass is lush and the animals don't need hay, so hay doesn't sell too well in summer. To sell hay in the winter at the highest price, you must first stack it in the hot, dusty barn in the summer, remove it by hand in the winter, transport it through snow and mud, and restack it in someone else's barn. Handling hay is not easy for anyone, and especially not for this part-time farmer and his family.

Finding Fruit

We discovered fruit almost by accident. We had grown strawberries for ourselves for several years and had some extras that we found we could sell quite easily. Other fruits also seemed to have a ready market. People came to us without a lot of urging, and we saw it as something we could do that could bring in a return. Nevertheless, we couldn't plant fruit trees on the property down the road because, in that neighborhood, security could be a problem. (We could grow the fruit without problems, but we might have more "help" than we wanted picking it—and that help would be there when we weren't.)

That's when we had the opportunity to buy 22 acres adjoining our original 5. It, too, had grown up in trees and brush, but this time I owned a chain saw, better tractors, and a Bush Hog rotary mower. Clearing the land took time, but not nearly as long as had our two previous pieces of ground. We expanded our pastures and our tree plantings. There is still some room for additional plantings if we think we need them, but that is not likely to continue unabated. One very big difference now is that we have a farm income to support this habit.

You might wonder about the name SweetAire. We needed to identify ourselves somehow, but didn't have a good name. Many of the farms in our area have historical names passed down from generations gone by, but our farm was our creation, and it didn't come with a name already attached.

We talked about this in our family, and really didn't have any good ideas. We considered naming it Waggin'-Tail Farm for a while, based on the many animals that we had, but with our new emphasis on fruit, that name wasn't appropriate. The name SweetAire Farm was inspired partly by a place in Maryland called Sweetair, which Cathy always liked—partly for the sweet aromas of blossoms in the spring, partly for the music that the children played on their instruments, and partly as a sarcastic description of the smell of manure being spread.

There is an old joke that goes like this: "How can a farmer become worth a million dollars?" "He starts out being worth two million dollars." Farming is often a losing proposition. It is long, hard work that doesn't usually pay off monetarily. But farming is more than just money, and a farmer may be better off than others making ten times as much money.

Pigs and Pumpkins

By Nancy P. Weiss

Editor's Note: In 1978, Nancy Weiss wrote about chasing pigs that broke through a poorly constructed fence and other mishaps of part-time farm life in northeastern Connecticut, in the original version of this book. Here are her reflections of the early years on the farm.

New England farms conjure up images of Robert Frost poems with bending birches and stone walls between neighbors. Some farms in Connecticut closely resemble the popular impression. My husband and I owned an 87-acre farm in Pomfret, a small town in the northeast corner of Connecticut. Built in 1790, the farm hugs a hillside and is reached by traveling up a long dirt driveway lined with maple trees.

Come spring, we tapped the maple trees. While sap buckets are picturesque, at our farm we replaced them with plastic tubing that conducted the sap down the hill to collection vats. Maple syrup was but one of several sources of either revenue or crops we produced on the farm.

We made our choice to live on the farm deliberately and at considerable cost, in many respects. Jim's jobs took him to Hartford each day, a distance one way of 49 miles. In Hartford he worked as a stockbroker for a large investment firm and served as the state legislator for seven rural towns. The legislative district, based on population, is geographically one of the largest in the state.

Our personal experiences on the farm put Jim in closer touch with the problems of many of his constituents, who were either commercial or part-time farmers. Through farm ownership, we worked with various USDA agencies and participated in a number of programs. The insight we gained through this firsthand experience was far more educational than learning of the work of the agencies through committee hearings or constituent contact.

I worked for the Cooperative Extension as assistant director of the 4-H Program in Connecticut. Our lifestyle required that I commute at least 20 miles each day to the University of Connecticut campus in Storrs. Although I was brought up in a rural area, my family did not farm. My work with 4-H provided me with experiences and an appreciation of farm life that was reflected in our choice to live on a farm.

Some days the juxtaposition between our professions and our farm was amusing and quite perplexing to nonfarm people. One year during the legislative campaign, I often had to call a halt to planning meetings or sign-painting parties while the volunteers helped me chase our wandering pigs back into their pen. Poor fencing was the problem, and highly intelligent pigs made quick use of our inability to take the time to properly fence them. News reporters laughed in astonishment when we dashed out the door of the partially restored farmhouse to call the hogs back to their pen.

One hot summer day, our tea party was disrupted by the exhaust from a faulty diesel tractor stuck in the hay field. Early-fall mornings saw us up at dawn to spray the pumpkin patch, our first effort at commercial growing, only to find that the harvest, nearly the entire crop, had rotted on the vine from heavy rainfall for several weeks.

Our small, peppery Scotty dog killed all of the chickens one morning—all, that is, that were left following an attack by a ravenous fox. Finally, a pet billy goat died from eating nightshade after two weeks of hand-feeding him and $61 in veterinary bills.

On my farm, I had many disappointments, but we tended to stress the successes. A bumper crop of green beans at our roadside stand more than paid for the investment of seed, fertilizer, and plowing. An added boon was the sight of a freezer full of homegrown produce, which easily met our family requirement for vegetables, provided we didn't become too sick of green beans, broccoli, and snow peas.

Our land had long been depleted by intensive growing of corn and poor soil conservation practices. We slowly brought it back into good condition. An abandoned apple orchard bore much fruit after several Saturday mornings of pruning. The woodland, once scorned as useless, provided wood for three fireplaces and a woodstove.

We felt a great sense of satisfaction. In restoring our house and a small half-house (which the farm help had used for many years), we discovered beautiful hand-done stenciling, a wainscoted "keeping room"—a central room of the house—wide pine board floors, and a long-abandoned bake oven.

Although we had to invest more money than we had hoped, much of ourselves went into the farm. As we pried ourselves out of our cars after long days and long commutes from work, we felt a surge of energy that came straight from the land. Several winter evenings while we sat snow-bound in the house, we spent our time planning better and more efficient ways to use the buildings and improve the land.

The View from Golden Hill Farm

By Nancy P. Weiss

Editor's Note: Since the late 1970s, Nancy Weiss's family has matured and changed—and so has her land. In 2003, she retired from her full-time job as director of development and alumni affairs at the University of Connecticut College of Agriculture and Natural Resources. In this essay, she considers her land, and tells what she plans to do now.

Working in the light, cold rain of a late-spring evening in my flower garden allowed me the privacy and the time to think about twenty-six years of life on Golden Hill Farm. So much had once seemed possible and reasonable and so much had changed. The buildings still stand and while in better shape and covered with paint, they hold no animals, except for the persistent barn swallows that find any crack and sneak in to make their mud-covered nests.

The big barn, renovated just last year, no longer has cow stanchions or horse stalls. The paddock fences have been taken down, and the empty watering troughs stand as unattractive reminders of the rough tongues that licked their sides on hot August afternoons. The hay fields that gave the place its name, Golden Hill Farm, still slope majestically down to the stone wall along the road, but now the hay remains standing until late July, after the ground birds have stopped nesting. The efforts to cut close to the walls and keep the edges neat have been changed to leave a buffer of tall weeds and grasses for habitat.

When our daughters were younger, we had the requisite pony and several sheep. The pony, Flicka, grew lonely when the sheep were slaughtered for one last dinner party of meat from the farm. She moved to a friend's house to keep company with a miniature horse and never returned.

Guinea fowl, which we raised to keep down the ticks, provided endless amusement with their paranoid personalities. They treated every nuance of cloud change with hysterical screeches. When a fox carried off the last batch, it seemed wise to give up on feathered tick control measures.

A local farmer allowed his beef calves to summer with us for many years. Wild and unfriendly when they arrived, we would ply them with goodies and scratch their itchy foreheads until they created a minor stampede every time we drove in the yard.

With every passing year, we abandoned another agricultural effort. But it is still a farm in spirit. It has the bones of a farm and the look of a farm, if only an antique one. It is, in fact, really a nature preserve now. As our little town filled up with houses, we decided to preserve the place forever from development. Over and over we walked the fields and thought about selling the property to someone who would really farm it. We would keep a little piece for ourselves, build a modern, efficient house, and be warm (for once) in the winter. We just couldn't see where their new house would go without ruining the entire property, so in the end, we didn't do it.

An eccentric, visionary neighbor filled her dying years with a radical activity. She purchased every bit of available land in our neighborhood, burned down every building on each property, and let it go back to the wild. Her plan worked. Game birds, turkeys, coyotes, bobcats, muskrats, countless deer, and myriad songbirds returned to the area. Eventually we sold the large front fields and gave the woods to the Connecticut Audubon Society through an arrangement she orchestrated. Many of our neighbors did the same.

Hundred of acres are now preserved and open to the public for birdwatching. One of the trails goes through our former field and woods and sometimes a hiker pops up in my backyard, but the intrusion is friendly and brief.

As I pulled away the old mulch and planted my annual cutting garden bed, I thought about the National Public Radio story I heard on the way home about mad cow disease in Canada. Last year I drove home with tears streaming down my face while I listened to this report from Wales, about a family farm where all the cows were destroyed due to the insidious disease. I began to think about getting some animals again and raising them myself. We could step out of the cycle of corporate farms and move closer to the self-sufficiency I dreamed about in 1978.

We have a contract with an organic vegetable farmer, who brings his wares to our town every Tuesday.

I wonder how much it would cost to fence that paddock again. This time I'll do it right so the animals won't get out.

Attracting Wildlife to Your Land

You made the right decision. You bought a little extra land. You have managed the down payment, rebuilt the house, built or remodeled a barn, and planted the garden. Maybe you have renovated an old orchard or asked the state forester for advice about Christmas trees. Health officers helped with the water and sewage. All in all, you are pleased as you settle back on the porch for a cool drink and some peaceful thought.

A flicker of color catches your eye—it's a small bird frantically trying to catch a bug, which is just as frantically trying to hide. A burst of birdsong signals success, or is it failure? No matter. It sounds nice.

Human beings evolved in and with the natural world. As we become more urbanized, more complex, and have more social pressures, we seem to lose touch with nature. We tend to forget just how closely our lives are tied to the other living things around us. But—the evening call of a bobwhite, a visit from a friendly squirrel, or the romp of fox pups can ease you out of some very trying days.

All right! You're convinced. You have the land and you want wildlife. Which species? Where? How long? When? Why?

Even on small acreages an amazing variety of wildlife can be observed. True, you can't expect to have a herd of deer, elk, or a flock of turkeys, but your land may be part of their range and they may pay a visit. Too often newcomers to the country have high hopes for large numbers of wildlife, especially if they are willing to improve their land. But as with most things, you must follow certain rules.

RULE NUMBER 1—DON'T MOVE A ROCK OR A BRUSH PILE. Don't cut a tree, plant a bush, dig a hole, level a field, or build a fence, until you have surveyed the land and know what resources you have. Once you know your resources, write out a plan of what, when, and how you will proceed with development of the land. You will make fewer mistakes with a well-thought-out plan to guide you.

An example of this is a landowner who raised cecropia moths for heart research. This moth thrives on chokecherry trees, normally considered a worthless weed tree in old field land. By inventorying the number of trees, the landowner determined how many moths he could expect to produce and the amount of netting needed for their protection. At the same time, he was also improving nearby habitat for songbirds that normally prey on the moth larvae. By planning the resource he could have both moths and songbirds.

RULE NUMBER 2—READ, ASK, AND VISIT. Find out what wildlife should be on or around your land and what wildlife is actually there. Zoos; local, state, and national parks; game and fish commissions; universities; and some federal agricultural agencies will be able to give you habitat requirements for both existing and potential wildlife. A wealth of nature books and regional guides will teach you about the birds, animals, wildflowers, or trees and shrubs of your area.

Walk over your land. Do this often and during all seasons to see the many changes. It is amazing what new things you will observe on each trip.

Find places where you can sit quietly and watch. This is the best way to see what creatures share the land with you. Learn to look at things closely, such as scratch marks on a tree trunk. This will tell you what lives there or uses the area.

RULE NUMBER 3—MANAGE YOUR WILDLIFE. Certain animals may increase beyond the available habitat. They need to be harvested or they will destroy their habitat or develop into nuisances. Trees and shrubs may grow beyond the reach of wildlife and become troublesome as weeds. They will have to be controlled and managed.

Planning for wildlife is like planning for the rest of the land. Wildlife, whether a pygmy shrew or a moose, need food, water, and some kind of cover. They set up territories just like humans. These things are part of the animal's habitat. Each kind of wildlife requires specific items in the habitat. To add, maintain, or attract specific wildlife, you must know what they require. For example, woodcock need small, open fields where they can go through their mating ritual flights at dusk on a spring evening.

Male ducks guard their marsh territory on a log or mound that you could put out for their use. Songbirds such as bluebirds respond to certain size holes in birdhouses. This offers them protection from larger birds, like starlings, which are more competitive for homesites.

Food Needs

All wildlife requires some kind of food. Plant eaters—some insects, elk, deer, rabbits, and porcupines, to name a few—convert plant energy to protein and fats. Wildlife such as predatory insects, hawks, owls, bobcats, insect-eating birds, and grizzly bears feed on the plant eaters. This complex web of food links will collapse without plants.

Some wildlife feed on the ground, some in small shrubs. Some feed only on evergreens, while others subsist only on broadleaf trees. Plan your habitat so a wide diversity of seed, berry, and foliage plants will be available. This is especially true if you are planning for a variety of wildlife but no one kind in particular.

Many plants provide both food and cover. On the deserts, some plants provide food, cover, and water. Plant as many multipurpose types as possible.

You can provide food not ordinarily available on a year-round basis to wildlife (feeding your chickens to a fox doesn't count). Birdseed, suet (beef fat), peanut butter, nuts, fruits, bread, sugar water, and the like can be provided in winter. Birds can also be fed in summer; for example, honey water tubes bring hummingbirds to your home, where they can be more easily seen.

Biologists are divided on the benefits wildlife receive from supplemental feeding. For instance, many of them now discourage the feeding of ducks and other waterfowl, because they begin to rely on it and can become aggressive. If you do winter feed, you must continue because wildlife become dependent on the source. If it is sporadic or cut off, they suffer and may die in severe weather.

Water and Cover

The importance of water for wildlife depends somewhat on where you live. Water is vital and a major wildlife attractant on the deserts in Arizona. If the land is surrounded by streams, bayous, ponds, or springs, water development on the land is relatively unimportant. Some wildlife, such as kangaroo rats, need no free moisture at all. Others need a readily available and constant supply.

Water attracts all kinds of wildlife, from the truly aquatic species to those that just casually use it. A water situation can be as simple as a birdbath or a small backyard pool. It can be a permanent part of the landscape or a temporary pool built of plastic sheeting. Farm ponds or tanks are more permanent types of structures. Small marsh developments provide habitat for wetland-loving wildlife.

On the Andrew Wyeth-James land along the coast of Maine, a shallow 2-acre marsh created in old field land attracted about forty different species of seabirds and shorebirds. They came to wade, feed, drink, or bathe in fresh water.

All natural wetlands such as potholes, marshes, swamps, bogs, ponds, or streams need to be considered in planning your land. Again, this brings us back to the basic rules of meeting the wildlife's needs.

You need to consider two kinds of cover: natural and artificial (or man-made). Both have a place in the plan for wildlife. Natural cover provides resting, roosting, nesting, protection, and foraging areas. This part of wildlife habitat is easily managed by manipulating vegetation. Vegetation can be planted, pruned, thinned, or cleared, depending on what is needed.

Artificial or man-made cover can be brush piles, nesting boxes, piles of rock, birdhouses, log piles, and similar structures. These types of cover can be a valuable addition to natural cover and benefit many kinds of wildlife. Dispersal of different cover types is important. Cover is needed for protection of wildlife travel routes and for escape.

Developing a Plan

If you farm your land, what you do can be good or bad for wildlife. Windbreaks, field borders, hedgerows, and farmstead landscaping can all contribute cover and food. Some types of draining or the burning of fencerows and roadsides may destroy habitat. Land not being farmed can benefit from reseeding, mowing, or rotation harvesting of hardwood timber that is beneficial to deer and smaller mammals or birds.

In developing a plan for your land, consider the following:

SOILS. Obtain a soils map if possible. This will provide information on what types of vegetation can be grown, where drainage may be needed, or where to build a pond or marsh.

VEGETATION. Find out what acreage is in grassland, shrubland, wetland, or forest. Locate the different types on an aerial photograph or map. This will give you the pluses and minuses in what you need for good wildlife habitat.

You may be pleasantly surprised to find that you already have all the wildlife plants you need. The vegetation may only need pruning, fertilizing, mowing, or releasing from competition with other plants.

WATER. Determine the quantity of available water and how it might be used. Consider all the possibilities such as wildlife, recreation, irrigation, or personal use. The quality and location are important.

WILDLIFE. List the wildlife you know are on the land or that have potential to be there. Write out their habitat needs and cross-check these needs against the habitat you have already. Whatever are lacking, you need to manage or plant.

PLANTING PLAN. Plan the kinds of plants you will need and their location. Be sure they are hardy in your area and adapted to the soils on your land. Determine quantities needed, cultural practices required in establishment, and, most important, their use by the wildlife desired.

Don't try to do everything at one time. Spread the plantings out over a period of time. The seed and shrub catalogs look great during the winter, but many good plans fall by the wayside when the time to plant arrives.

Go to some places where these plants are already established, see what they look like, how the plants produce, and ask if this is what you really want.

When you help wildlife, you help yourself. The scraping of a cricket, the sight of songbirds, and the glimpse of a fox all tend to wring out the tension of modern living.

The Benefit of a Woodlot

When you have trees on your land, you have numerous advantages. A woodlot, or small forest, can be defined as a dense growth of trees on an acre or more, usually fewer than 100 acres. A few scattered trees cannot be construed as a forest. There must be many trees, each dependent on the others and all forming a community. This close association is important in order to maintain productivity through planned management.

Many owners cherish their forests for the multitude of intangible riches they yield, and accept the wood harvests as an extra benefit. Others consider the wood yield as the primary objective. Some landowners are blessed with an already established forest when they purchase their property, while others begin with bare land and "build" a forest by planting seedlings.

Establishing a plantation usually requires the planting of 400 to 1,000 seedlings on each acre. In ten years, these small trees should reach a height of 10 to 20 feet, depending on the species. At this point, their crowns should touch, and the entire plantation begins to appear as an established forest community.

Planning

No matter what stage of development a forest is in, you need a plan that describes what you'll do to make the forest healthy and productive. You need not be a forester to prepare this plan, but advice from one would be extremely helpful. Foresters are available through most Cooperative Extension programs, often free of charge, as well as private companies and universities.

The plan should list your objectives and a timetable for meeting them. Your plan should include a map and an inventory of trees by species and size. The plan should not be absolutely rigid, as you will need to periodically amend and change it to fit circumstances.

Renewal

You should provide for continuous replacement of woodlot trees. Plant trees with special tools or machines. State forestry organizations offer seedlings at nominal costs; you might also check with private nurseries. One person can plant about 500 trees in an eight-hour day, but with a planting machine that number jumps to nearly 10,000 trees.

Most forests can be restocked naturally. This system requires careful planning. For instance, many species need an abundance of light to develop, and won't survive beneath larger trees. Others prosper in the understory. Because of such critical differences, you must create conditions that will stimulate the well-being of each species.

Improvement

Once a woodlot is established, it needs attention and care. If it turns into a neglected area, its productivity and usefulness will be greatly reduced, and sometimes it becomes a liability. All woodlots don't require the same care. Various needs depend upon the species of trees, markets, past treatment, age of the forest, and your personal objectives.

As trees increase in size, they need more space, moisture, nutrients, and light, resulting in the need to thin them. A pine forest may begin with 1,000 trees per acre, and after several thinnings at ten-year intervals, the final harvest may yield only 100 large, high-quality trees. If you don't thin such a forest, many trees would die, and the final harvest would yield several hundred small, low-quality trees.

Trees removed in thinnings can usually be sold at a profit, unless they are quite small. Nevertheless, you will profit by having higher-quality trees to market in subsequent sales.

All trees don't develop into beautiful specimens. Some grow crooked, others decay and become hollow, and still others suffer root or top damage. Foresters call these trees "culls." You must remove them. They are like weeds in the garden. Leave a cull if squirrels or other wildlife have moved inside. Leave a flowering species, such as a dogwood, that you like, even though it will never yield a commercial product.

Knotty trees are low in quality. Avoid this problem by pruning off limbs from the main stem at an early age. Never remove the limbs from more than half the total height of the tree and only from chosen "crop trees"—trees that will form the final harvest.

When growing a very valuable species such as black walnut, where you want maximum diameter growth, don't allow the stand of trees to become dense enough to cause natural pruning. Do the pruning manually.

Cutting Trees

One of the most crucial periods in the life of a woodlot is when trees are selected for harvest. The improper choice of trees can result in a substantial reduction in future productivity. Forests have been completely destroyed by harvesting the wrong trees.

Maturity is only one reason to harvest a tree. Consider also how other trees nearby will be established. It is best to operate under the guidance of a professional forester when selecting trees for harvest.

Protection

Forests can be decimated by a number of enemies, and as the owner you must be on guard at all times. The greatest and most feared enemy of the forest is fire. A single catastrophe can kill trees, scar others, damage the soil, and even bring about the invasion of insects and diseases. Many years—sometimes as many as one hundred—are often required to recuperate from just one fire.

Prevention is the key to protection from this nemesis. Many forest fires start from carelessness with trash burning, campfires, and smoking. Become familiar with the local forest fire control organization. Knowing how it operates, the services it offers, and how to obtain help in case of fire can be important.

Small woodlots will receive some protection from cleared firebreaks that are at least 8 feet wide. Keep firefighting tools nearby. With a confident knowledge of how to use these tools, you can control small fires before they cause damage.

Although in some portions of the country controlled grazing by domestic livestock is permissible, intensive grazing is a very harmful practice in most areas. And, even though it doesn't destroy a forest as quickly as fire, grazing can accomplish the same disastrous results. It is sometimes referred to as "creeping devastation."

In general, woodlots are a poor place for livestock. The forage is sparse and of poor quality. Livestock not only trample and browse young trees and injure the larger ones, but also compact the soil. This compaction reduces tree growth and increases water runoff during heavy rains, thus promoting soil erosion. Remember—cows make poor foresters!

Insects attack trees. A poorly managed forest of weak trees is usually more susceptible to insect damage than a well-managed one composed of vigorous trees. Insects attack by defoliating, boring into the twigs and roots, and by girdling the stems. Some insects also attack the fruit and seeds, which reduces the reproductive capacity of the trees. Close observation and the use of early control measures is the best way to prevent excessive insect damage.

Many diseases attack trees. Some cause only minor damage, while others have been so harmful that certain species of trees have been almost eliminated from American forests. The chestnut blight and Dutch elm disease are examples of such devastating diseases. Early detection and prompt control are important.

Selling the Trees

Seek help from foresters to maximize your profit from selling trees. The first step in marketing is to select the trees to be cut and to mark them, preferably with paint. Tally each tree by species, size, product, and volume.

Determine current prices for similar quality material. Most states have periodic price reports available to the public, free of charge. Properly informed, you are ready to bargain with buyers. A common practice is to obtain bids from several buyers before selling.

After deciding on a buyer, a contract should be prepared that provides penalties for violating certain provisions. Several items of concern are: the amount, method, and time of payment, logging damages, cutting unmarked trees, and time allowed for removal.

If you have the necessary equipment and the basic knowledge, you can harvest your own trees. So, instead of selling them as they stand, you can log them and sell at the roadside or mill. You can make more money this way.

Since you will often need lumber as a homesteader, it's an added benefit to get it from your own trees. Have your harvested trees cut into lumber at a mill. Drying and planing services may also be available if you desire finished lumber.

Never hurry when marketing your forest products. It took many years to grow them, so spend a few extra days to ensure the best return.

Side Benefits

Woodlots also produce:

GAME. Occasionally, an owner will sell hunting privileges to his woodlot, thereby realizing an annual income. Many, though, prefer to manage for game so they will have their own private hunting preserves. Others prefer to establish and perpetuate an abundance of wildlife just to observe and not hunt.

Even if you believe you don't like hunting, consider that wildlife may become too abundant and subsequently cause damage to the forest. Ask any rural woodlot owner in the Northeast, where the white-tailed deer has made a comeback of unbelievable proportion in the last fifty years. Deer eat everything young in a forest, including young trees. Learn the game regulations in your state.

Nuts. Black walnut, pecan, and hickory trees can provide a decent harvest. Some forest-grown walnut trees produce more than 100 pounds of nuts in a single year.

Maple syrup. In the Northeast the sugar maple tree abounds. The landowner who has a forest containing this species has the option of operating a maple syrup production business. Tap the trees for sap in early spring, then boil it down. About 40 gallons of sap are required to produce one gallon of syrup. The content of sugar in the sap varies from tree to tree; therefore, sap from "sweet" trees requires less boiling, and fewer gallons of sap are needed per gallon of syrup.

Watershed protection. Woodlots not only yield clean water for personal and wildlife use, but also prevent soil loss and flooding by reducing water runoff. In certain critical areas, the forest's principal purpose is watershed protection. In all instances, it is of primary importance.

Recreation. A forest provides a place for camping, hiking, picnicking, and hunting. The building of hiking trails, camping sites, and picnicking areas within even a small woodlot can result in many hours of pleasure for you, the owner, and others. When utilizing a forest for recreation, take care to protect the trees from damage.

Firewood. With energy conservation a necessity, this source of heat can be of great value to the owner. If your supply exceeds your needs, ready market exists for the surplus. Use culls and otherwise undesirable trees for use as firewood.

PART 5

Growing and Raising: How to Do It

Fruit Trees

Tending an orchard is intensive work, but you can produce good fruit on very small pieces of land. If you have as much as 20 acres, you can make a good living. It's important to plant in the right location and invest in preparing the land, getting equipment, and choosing the best trees. You will see little return for the first three to seven years.

Be efficient and don't neglect controlling weeds, pruning, controlling pests, and fertilizing. Fruit trees show signs of your neglect after

CONSIDERATIONS BEFORE ESTABLISHING AN ORCHARD

- Need for training of operator
- Availability of advice
- Types of fruits grown in the area

- Rainfall distribution, needs for irrigation
- Winter temperatures
- Temperatures during blossoming period control
- Marketing possibilities

- Availability of labor for harvesting and operations
- Variety mix for labor distribution

- Variety mix for orderly marketing
- Variety mix for pollination requirements

- Machinery needs
- Need for storage facilities
- Possible rental of machinery or custom operators
- Site: air drainage
- Site: soil drainage

- Soil preparation: nutritional requirements of fruit types
- Purchasing, storage, and handling of other pesticides
- Human factor—tolerance to pick-your-own operation, pesticide regulations, etc.

one season. When the trees do produce, if you have a number of them, you'll need people to help pick. The requirement for, and investment in, machinery and supplies will be about the same whether you grow 2 or 20 acres of fruit. If the machinery can be adapted to other enterprises, shared with other orchardists, or rented, or if custom services are readily available, a smaller orchard unit may be practical.

As with other perennial crops, fruit trees can suffer damage in certain areas of the United States. The potential problems include: the cold; spring frosts leading to inadequate fruit set; damage from hail, wind, and rain; inadequate moisture or too cool temperatures leading to small fruit; or fruit that doesn't ripen at the right time.

The Great Plains areas generally produce poor fruit. It is too cold, and not enough rain falls there in the summer. Most northern states are subject to occasional winter freezes and spring frosts, which reduce the chances of annual crops. In the deep South, certain fruits and specific varieties of other fruits may not receive sufficient winter-chilling to blossom and put forth leaves properly.

Look for successful orchards in your area. If there are none, that isn't necessarily a statement that orchards wouldn't do well. It could mean that the orchards became house subdivisions. Only consider an orchard if

SUSCEPTIBILITY OF TREE FRUIT CROPS TO ENVIRONMENTAL HAZARDS

Hazard	Apricot	Apple	Cherry (sweet)	Nectarine	Peach	Pear	Plum
Spring frosts	xxx	x	xx	xx	xx	xx	xxx
Winter freezes	xx	x	x	xx	xx	x	xxx
Insufficient chilling	x		xxx	xx	xx		x
Brown rot	xx		xxx	xxx	xx		x
Rain cracking	x		xxx	xx	x		x
Bird damage	xx		xxx	x	x	x	xx
Tree borers	xx		xx	xx	xx		xx
Hail	xx	xx	xx	xx	xx	xx	xx
Excessively high temperature		x			x		x
Insufficient moisture	x	x	xx	x	x	x	x
Poor soil drainage	xxx	x	xx	xx	xx	xx	xxx
Bacterial diseases	xx	x	xx	x	x	xxx	xx

x = slightly susceptible   xx = moderately susceptible   xxx = highly susceptible you have a suitable piece of land. For example, in northern areas, a site on the windward side of a large lake may escape low temperature extremes. In general, what you want is a sandy loam soil. Clay soils do not drain well, and fruit trees don't like "wet feet."

Check several years of records of the low temperatures as close to the proposed site as possible. If winter temperatures often drop to -22 degrees Fahrenheit, don't try to grow fruit. If temperatures fall below -13 degrees in some years, expect peaches, plums, nectarines, and apricots to suffer. Seek land where the temperature seldom goes below -4 degrees. Of course, that encompasses the majority of the country. Fruit trees are pretty hardy.

If you want to sell your fruit, the marketing possibilities may determine what kinds of fruit you grow. You may live near commercial wholesale outlets for fresh fruit, canning, freezing, or other processing uses. If you live near large population centers, consider setting up a fruit stand, selling regularly at the burgeoning farmers' markets, or running a pick-your-own operation. Don't underestimate the economic value of allowing your customers to wander among your bounty. Such outings have become extremely popular.

Also, you must decide whether to grow only one fruit type or a mixture. For fruit stand or pick-your-own operations, it's better to grow several varieties. The customers will want something ready to harvest throughout the warm months. If you choose to grow one type of fruit, you still can harvest all at once and sell to wholesalers. You lower the risk of failure if you grow several varieties, unless your climate is cold enough that the hardy apple tree is the best choice.

If you plan to get the necessary air drainage for frost control with steep slopes, you must control erosion with contour planting and sod maintenance.

It's important to know the previous use of the land. Crops such as tomatoes, potatoes, and melons favor buildup of harmful nematodes and wilt fungi. Newly cleared forestland may be infested with root-rotting organisms that also attack some fruit species. Contact your local state university cooperative extension office for help in testing the soil or preparing it for fruit.

Full-Sized Trees, or Dwarf?

Decide how to set up the orchard. You need far fewer standard-size trees than dwarf trees in a given acreage, but standard-size trees have the disadvantage of producing fruit inaccessible except from a tall ladder. Spraying, if you do it, and pruning and thinning, which you must do, all require more powerful equipment to reach the larger trees. Standard-size trees take a few more years to produce fruit, but dwarf trees grow less vigorously. Dwarf and semi-dwarf trees are easy to tend, but they are most hardy if you choose apple varieties.

The investment in trees multiplies as smaller trees are planted closer together. Trees of the dwarfing stocks grow differently in different areas, so follow local recommendations for planting distances. Poorer soils produce smaller trees, and thus more trees can be grown per acre. In general, dwarf trees require better soil and better air drainage than standard trees, and they produce fruit earlier.

Most peach and nectarine varieties are self-fertile and may be planted in solid blocks. For the other fruit trees, you will need a pollenizer variety. Most fruit nursery catalogs indicate satisfactory pollenizers for specific varieties.

A rule of thumb for providing pollenizers is every third tree in every third row, which exposes one side of each main-variety tree to a pollenizer tree. For ease of handling, it may be more feasible to interspace two to four rows of the main variety with one or two of the pollenizer variety. Usually both varieties can be commercially acceptable and can mature at the same time—or at different times if a spread of maturity is preferred.

With peaches, it is now possible to have good adapted varieties ripening at weekly or even semi-weekly intervals for almost four months in areas suitable for this fruit. The ripening period of the other fruit types is shorter, but a sequence of good varieties is possible.

Give thought to the ripening sequences if you grow several types of fruit. Apricots and cherries ripen earlier than the other fruits. However, pear ripening coincides with that of later peach or early apple varieties. Decide which fruit crop will be ripening in each period, unless you can hire help to harvest.

It might seem tempting to plant vegetables or other crops between the fruit trees before they bear fruit. Don't do this, because most vegetable crops probably require very different conditions than the fruit trees. Also, machinery used for the trees does not adapt to the other crops, and it won't be worth the cost to buy specialized equipment just for three to five years' worth of vegetables while you wait for the fruit to bear. Trying to combine both fruit and vegetable crops may result in neither crop receiving the care it needs at the appropriate time, or in one crop being neglected. Give fruits and vegetables their own separate sites.

Invest Adequately

Do not consider growing fruit unless you can provide for adequate equipment. You will want a tractor for mowing, cultivation, and spraying. A rotary mower or disk controls weeds between the rows. The use of dwarf trees and close planting distances may require staking or wire trellises, and will call for smaller equipment.

You probably will need a tractor forklift and harvesting wagon. Make arrangements for harvesting buckets, boxes, or bins, and ladders, if the trees are standard sized. Cold storage facilities are almost a necessity if you plan to operate a fruit stand or if you want to store fruit into the winter.

Dealing with Pests

If you grow fruit trees, you must learn about roughly 40 pests and diseases that threaten them. Fruit trees are terribly susceptible to pests, but the heavy pesticide use that started after World War II hasn't withstood the test of time. Doing nothing and calling it the organic approach is a bad idea in an orchard. You can get away with it to an extent with other produce, but fruit trees need a lot of babying. To ignore them almost certainly would yield mottled, diseased, bug-filled fruits. Yet any orchard manager today must be careful about chemicals. The federal government has outlawed several chemicals that were in common use when this book first came out. To control fruit tree pests and diseases with chemicals, be sure you know what you are using and how to apply it. If you do not use any chemicals, be prepared to work hard to keep insects from taking over. Most pesticides in legal use break down before the apple ripens, but residues remain and it's beyond the scope of this chapter to summarize the various arguments on whether those are dangerous. Certainly a percentage of your fruit buyers will think so.

The solution for many growers is to practice integrated pest management, the practice that combines physical control like mowing with judicious use of chemicals considered safer because they break down more quickly. You can use hanging apple maggot traps, for instance. These cut the number of maggots without depositing chemicals on the fruit. New research has shown that biological controls can work. For example, researchers found a predatory mite feeding on the mites that harm apples in orchards without chemicals. Read about the latest research on pest control by searching the USDA's Research, Education, and Economics Information Center at http://www.reeis.usda.gov. The search function can yield a thousand documents difficult to sort, so a Google search for REEIS and orchards might take less time.

An Established Orchard?

Should you consider buying or leasing a small established orchard or a farm with a small orchard? Many factors must be considered and expert advice sought before you make such a decision:

- Analyze the prospects. Is the orchard in good condition and, if so, are the trees young enough to remain productive? Are the varieties commercially desirable? Is the mix of types and varieties suitable to the operation planned?
- Determine the production and profit history of the orchard. Are the necessary outlets available for the type of operation planned?
- If the orchard is not in good condition, what is the reason? Are there indications that the site or soil conditions are not suitable?
- Have the trees been neglected and, if so, can the orchard be renovated, or must it be replaced? If replacement seems practical, are there replant problems such as arsenic in the soil from previous sprays, high nematode populations, root-rotting organisms, viruses, or nutrient deficiency indications?
- You should also determine whether storage facilities are available or will be needed. Unless you can reasonably expect a profit from the management you propose for the orchard site, don't buy it. In most cases, it will be better to invest your labor in establishing an orchard suited to your own situation.

Planting and Management

The number of fruit tree varieties will amaze you. Research which you want and what will do well on your land. One excellent place to start is an online catalog of sorts operated by the University of Minnesota Library at http://plantinfo.umn.edu. It will lead you to growers that sell rootstock in your region.

Plan for the soil management well in advance. If high populations of nematodes or root-rotting organisms are indicated by previous cropping or by soil tests, you might need to fumigate. Fumigation is expensive, but should reduce tree loss and also provide weed control the first year as a bonus. When planting peaches, it's possible to use resistant rootstocks if root-knot nematodes are high.

Certain rootstocks, such as Stockton Morello for cherries, and peach-almond hybrids for peaches and plums, will alleviate excessive moisture or high alkaline conditions.

A green-manure crop the year prior to planting is a good idea. If the soil pH (which indicates acidity or alkalinity) is lower than 6 to 6.5, lime the soil to supply calcium as well as to correct the pH. Be careful in selecting varieties. Check on the recommendations of your county Extension agent and of experiment stations with conditions similar to yours. Time spent with nursery catalogs or nursery staff specializing with fruit types in your general area will help you select adapted varieties.

Varieties differ widely in their adaptation, so don't make a large planting of an untried variety. Test a few trees first, if the variety is new to your area.

You can propagate your own trees, but this involves growing root-stocks, finding a "clean" source of budwood, making the propagation, maintaining the nursery, and waiting usually two years to have trees comparable to those of commercial nurseries. Instead, select a reputable nursery and order medium-sized trees of adapted varieties. Order not only the variety you want but the rootstock that seems appropriate for your conditions.

Many of the catalogs give useful information on planting distances, number of trees needed per acre for different planting distances, effective pollenizers, planting directions, and training and pruning advice.

NUMBER OF TREES PER ACRE AND YEARS REQUIRED TO REACH FULL PRODUCTION FOR VARIOUS TREE FRUIT TYPES

Fruit type	Density	Number of trees per acre	Years to reach full production
Apricot		75–200	5–7
Apple	Low	Less than 100	15–20
	Moderate	100–200	9–15
	High	200–500	6–9
	Ultrahigh	More than 500	4–7
Cherry (sweet)		75–150	9–15
Nectarine & Peach		100–200	4–6
Pear	Moderate	100–200	6–8
	High	200–500	4–6
Plum & Prune		100–250	6–8
Sour Cherry		100–200	5–7

Follow the customary planting time and procedure for your area. Where winters are mild, fall planting gives the roots time to become well established before growth starts. Have the nursery store your trees until you can plant, unless you have facilities to store them properly.

Avoid drying out the roots during planting. Usually it is best to plant them directly from a barrel of water into the soil. Water them in if the soil is dry. Adding some water will ensure better contact of roots and soil.

The training system starts with planting. The planting distance determines how much space a tree can occupy without undue shading.

Check with Extension service agents or in planting guides for detailed pruning instructions. Usually the newly planted trees are topped at 2 to 3 feet, and each standard-sized tree has two to four main leaders. High-density trees require more specialized pruning. In either case, keep pruning to the minimum necessary to develop the main leaders and to avoid overlapping limbs.

Well-exposed wood that is not part of the permanent tree structure can be saved for early fruit production. Prune in the summer as much as possible, to prevent delay in fruiting.

Time fruit harvesting to place the highest-quality product possible in the hands of the consumer. Appearance matters to customers. Production of high-quality fruit requires picking at proper maturity. There are many maturity indices, but development of the undercolor and softening of the flesh are the most reliable.

The proper stage depends somewhat on how you plan to sell the fruit. Seek local advice. For example, fruit to be shipped to distant markets undergoes some ripening in transit and handling and, therefore, cannot be picked fully tree-ripened.

Apples may be stored for varying periods in refrigerated storage at about 31 to 34 degrees Fahrenheit, depending on the variety. Pears usually are stored anywhere from a few weeks to four months, depending on variety. The other fruits usually lose quality after storage for two weeks or less. Modified atmosphere storage can extend the period of storage. (The atmosphere is modified to 2 percent oxygen, about 3 to 5 percent carbon dioxide, with the remainder nitrogen.)

Always look for good advice from your county Extension office, nearby experiment stations, commercial and amateur growers in the area, and reputable nursery operators. Helpful handbooks are available in most fruit-producing areas for the adapted fruits.

Grapes

The grape is a versatile fruit. For dessert use, there is a wide range of flavors—fruity, spicy, muscat—and of textures—crisp, melting, and juicy. Seedless varieties come in the same range of colors as seeded. Grapes make jelly or breakfast juice for use throughout the year. With some additional effort, wine from your own vines can enhance your meals.

You can sell grapes for all of the above purposes (except as wine, which is strictly controlled by state and federal laws). This crop can be a profitable, though sometimes risky, source of income. Grape growing has been tried at some time during the last 400 hundred years in almost every area of this country. Grapes are produced commercially where the climate is favorable, diseases are successfully controlled, and vines survive well enough to pay back the costs of a vineyard.

In established grape areas, the competition for markets is greater than in areas where more care and skill are required or where the risk of failure is greater from frosts or rots due to untimely rains. Modern pesticides and the use of varieties adapted to particular regions have extended the areas where grapes may be profitable.

If you can accept an occasional crop failure and plant those varieties best suited to your locality, rather than attempting to grow the "premium" varieties of the better commercial areas, you can grow grapes almost anywhere where frost subsides for 140 days or more.

There are too many varieties, climates, and judgments of quality to cover this subject in detail here. A general listing of types of grapes will help you narrow the list of possible choices. (After you do this, contact your county agricultural or Extension agent for more specific ideas.)

Vinifera (European or California types). Among these varieties are found the highest-quality grapes for table, raisin, and wine use. Unfortunately, they are very susceptible to diseases, and some are damaged both by winter cold and by warm spells in the winter.

A few small plantings have been made outside the recognized West Coast vinifera areas, but they are considered too risky to recommend for general commercial plantings.

American. These are the hardy, disease-tolerant varieties. Most are derived in some degree from the wild American fox grape *(Vitis labrusca)*. Well-known examples are Concord, Delaware, and Niagara. Several modern varieties in this group are seedless and well suited for table use.

Hybrids. These include the so-called French hybrids developed in France over the past century, primarily for wine. They are more neutral in flavor than most American varieties, and a few are well adapted for table use. The hybrids are crosses between European grape varieties and various native American species. They were selected for the fruit quality of their European ancestors and the disease and insect tolerance of their American ancestors.

Both traits vary widely, from nearly wild with excellent resistance to pests, to high quality with modest resistance. As a group, hybrids are much easier to protect against the many pests than are the vinifera varieties.

Ripening of hybrids varies widely, and there are varieties suited to most growing seasons. Also in this group are several of the more recently introduced wine and table grapes that originated in the United States.

Southeast grapes. Pierce's disease is widespread in the southeast United States. It limits the survival of vinifera and most of the American and hybrid varieties. Muscadines (Scuppernong types) are tolerant, as are a few of the older American varieties and some recent American varieties originated in Florida.

Whether rootstocks are necessary, and if so, which one to use, is especially confusing to the beginner. For most U.S. vineyards, only the vinifera varieties may need to be grafted. A rootstock enables you to grow a variety under conditions where the own-rooted (nongrafted) vine might fail. Reasons for failure include phylloxera root louse damage (on heavy soils) and nematode damage (on lighter soils, especially when replanting a vineyard).

Certain rootstock varieties may increase vigor, permit good growth on lime soils, reduce damage from drought-prone or wet soils, or tolerate certain soilborne diseases. Rootstocks are advisable only when one of these particular situations has been found repeatedly in the area. The rootstock variety chosen is of the group that best corrects the situation.

Before planting grapes you must know what is to be done with the crop. A quarter acre (100 vines) can produce a ton of fruit (50 to 60 bushels). This amount would make 100 to 200 gallons of juice or wine.

The date of harvest of a variety can vary a week or two from year to year. When the fruit is ready or if the weather turns wet, picking cannot be put off, and once the fruit is picked it must be used promptly. A ton or more of grapes with no visible market can lead to a state of panic.

Except in areas where the grape supply is at or near saturation, there are several ways to market the crop. Successful marketing depends on the correct choice of varieties made when planting the vineyard.

When establishing the vineyard, you need to consider the time and expense of these elements: vines, trellis posts, wire, pesticides, and labor. It takes approximately three years for the vines to begin to produce. Labor requirements for weed control, pruning and training of vines, and pest control are high. Many of the specific chores must be done at the right time or major (and expensive) problems can develop.

Successful grape production always depends on keeping ahead of the problems. You must take preventive steps against weeds, insects, and disease. Once any of the pests build up in a vineyard, control is more costly and risk of financial loss greater.

Recognizing and staying ahead of the problems extends to all phases of grape growing. For example, replace weak posts promptly, or the entire row may fall over. Other problems include rabbits that prune off young vines, deer that seem to prefer tender grape shoots to almost anything else, and birds that flock to the vineyard before the grapes are ripe enough for you to enjoy them. You must consider adding the cost of fences or netting to your estimates for growing grapes.

Where to Plant

To be productive, grapevines require full sunlight. Nearby trees, even when they do not shade the vines, compete for moisture and provide birds with a perch from which to invade the vineyard. Wet soils are not good for grapes. At least 3 feet of medium to heavy soil or 5 feet of sand should be available. Shallower soils reduce vigor and size of crop.

Grapes are adapted to a wide variety of soil types. The low fertility of light or poor soils can be corrected with fertilizers. Deep, rich soils can result in overly vigorous vines that have poor clusters, mature later, and are of lower quality for wine.

The extremely steep hillside vineyards above the Rhine are picturesque and the slope does increase the sunlight and warmth reaching the vines, but only the high prices received for these special grapes justify the labor involved.

Less extreme slopes may be practical in areas where air drainage gives protection against frosts after the tender vine shoots have begun to grow. In most areas of this country, nearly level to rolling ground is more practical for grapes. Avoid frost pockets, those areas where cold air tends to collect at the bottom of slopes.

Spacing of vines is determined as much by the equipment to be used as by the varieties grown. Ten to 12 feet between rows is recommended so that tractors and sprayers can travel between the vines. A closer spacing is possible if you use smaller equipment, but the actual yield per acre does not increase in proportion to the number of vines. In humid areas where diseases are prevalent, close spacing of rows slows the drying of leaves and fruit and makes disease control more difficult.

Spacing between vines within each row is determined by the vigor of the variety. You want the vine foliage to at least meet between the vines so that maximum use is made of sunlight. Some overlap of foliage is usual and expected. Vigorous vines will fill the trellis with foliage when spaced at 9- to 10-foot intervals. Some of the less vigorous hybrids need a spacing of 6 feet between vines. Fruit yield is based on foliage exposed to the sun, so as long as the trellis is filled, optimum yield can be expected. Increasing the number of vines does not increase the yield.

The Size of the Crop

Don't ignore the crucial fact that a vineyard produces only so much fruit and no more—if you try to coax more out of the vines, the fruit won't be as good.

Sunshine ripens the grapes. Under ideal conditions, a long growing season with no cloudy days or rain at the end, a 1-acre vineyard (regardless of the number of vines) can produce about 15 tons of properly ripened grapes per year. In a shorter growing season, or when clouds or rain reduce the available sunshine, the maximum yield may be only a quarter of that. If you permit the vines to bear too much fruit, the sugar content will be lower, vines may be damaged, and fewer fruitful buds will be produced so that during the following year the yield will be much smaller.

Varieties and the number of vines to plant for your own use are determined by your goals. For juice, ten Concord vines set at 8 to 10 feet apart will usually provide 50 quarts of grape juice each year.

For table use, each vine will produce 5 to 15 pounds of fruit. One or two vines each of ten varieties chosen to ripen over a period of one and a half to two months will usually produce enough for lavish home consumption and gift baskets for friends.

For wine, $1/4$ acre is the maximum size for a home winemaker operating under the federal permit limit of 200 gallons per year. The size can be scaled down to suit your goals.

Varieties grown will depend on the type of wine desired. Some varieties are acceptable for either table or wine use, and the wine grapes do make interestingly different unfermented juice or jelly.

Running a Winery

The 2000–2010 decade will be known as the decade of small wineries booming. In New York State, for instance, more wineries have been opened since 2000 than the sum of all opened in that state ever—86 opened in the decade of 2000–2010. Most of the wineries in operation in New York State sell their wine at farmers' markets rather than direct to liquor stores. In Oregon, wineries' acreage and number of wineries are increasing. Washington State's grape production was at an all-time high of 145,000 tons in 2008, a 14-percent increase over the previous year.

The Alcohol and Tobacco Tax and Trade Bureau offers clear overviews at www.ttb.gov about running wineries. There are four types of wineries:

- Bonded wineries, for those who want to own their own winery and produce wine for sale.
- Alternating proprietors, wine producers who want to make wine to sell but do not want to run their own winery. Instead, they would share a winery facility with others and be taxed accordingly.
- Wholesalers, also known as a "custom crush clients," might grow grapes or buy them but would like a wine maker to make it for them. They are taxed as wholesalers.
- Bonded wine cellars are warehouses to store, blend, or bottle wine.

Laws on shipping wine from your farm to customers differ from state to state. For an overview see www.freethegrapes.org/state_laws.html.

USDA grants boost small wineries

Ed Boyce and Sara O'Herron are former management consultants who have long been wine enthusiasts. They have a home in Silver Spring and commute daily to their 146 acre farm in Mount Airy to care for 22.5 acres of wine grapes. With the help of a USDA Rural Development grant a few years ago, Boyce and O'Herron received $199,553 from the USDA Rural Development Value Added Producer Grant Program. They used the funds to help pay for working capital expenses that include barrels, bottles, wine making supplies, and support to launch their marketing campaign.

Other Maryland vineyards that have received these grants in recent years include Diamondback Wine ($49,872), Elk Run Vineyards ($263,000), and Mark Cascia Vineyards ($99,856). Small wine operations in Virginia and Michigan also received grants—along with farmers around the United States producing cheese, canola oil, beef, vegetables, berries, and more.

Value-added products are created when a producer takes an agricultural commodity, like milk or vegetables, and processes or prepares it in a way that increases value to consumers.

Black Ankle Vineyards, like many new small farms, is a sideline venture for its owners, who bought the farm in 2002, planted their first vines in 2003, and began planting additional vines in 2004.

"We truly believe that the geology, soils, and climate of the Maryland Piedmont are capable of producing world-class wines, and Black Ankle Vineyards is committed to turning that belief into reality," said co-owner Sara O'Herron.

Selling Grapes

Successful marketing depends on planting the right varieties for both your climate and the intended end-use of your customers. The quantities you can expect to sell to each customer vary greatly. Roadside stands are a popular marketing method, and grapes make an excellent addition to other fruits and vegetables in attracting customers. Three classes of grapes can be marketed:

- **Table grapes.** Top-quality, well-ripened clusters are required. Individual sales are small, ranging from 1 to 5 pounds, and some sort of container is required to prevent damage to fruit. Price per pound is high.

  The great range of varieties can be used advantageously. Early- to late-ripening plus cold-stored clusters of some tougher varieties can further extend the season. Possible flavors include Concord, muscats, and neutral grapes. Seedless varieties will be especially popular.

- **Juice and jelly grapes.** The flavor generally expected by a customer is the Concord-type of fruity fox grape. Quantities for each sale may range from a half to a couple of bushels. Smaller clusters and a trace of imperfect berries are acceptable. Some not-quite-ripe berries with their higher pectin content are even desirable for the jelly maker.

  The neutral or muscat-flavored grapes make interesting juice and jelly. The offer of a sample of the finished product is a good way to convince a prospective customer.

- **Wine grapes.** Some people enjoy home winemaking but don't have the space or time to grow their own grapes. The quantities you can expect to sell to each customer range from one or two bushels to several hundred pounds.

The wine quality from each variety is determined by proper ripeness of the grapes. While perfect fruit is desirable, a few damaged berries can be removed when the customer prepares fruit for fermenting. Aim for sugar levels of 20 to 22 percent.

Pick-Your-Own

Eliminate the cost of harvesting, except for some supervision in the vineyard, by letting your customers pick the grapes. The customer reduces his cash outlay and is able to do some trimming-out of defective fruit while picking.

If you decide to specialize in wine grapes, there are several ways to make the marketing more attractive to home winemakers. After the fruit is in the basket, the home winemaker is faced with the problem of crushing and pressing white grapes or stemming and crushing the red grapes. Several growers, who may themselves be home winemakers, have purchased small commercial stemmer-crushers and presses and either lend or rent these to their customers.

Some growers offer (for a price) instructions in winemaking, guidance and equipment for sugar and acid determination, and suggestions for the best handling of each batch of wine.

Other Markets

You may be able to sell grapes to supermarkets, local grocery stores, farmers' markets, or to stores that cater to home winemakers. You may be able to sell your entire crop to a commercial winery. Prior arrangements—even a contract—on varieties, acceptable sugar levels, and price should be made before planting. The price per ton received will be lower, but so will be the cost of marketing.

Planting Strategies

Because grapes are expensive to grow and take several years to reach maturity, consider several strategies in planning a vineyard. If you are fortunate enough to know exactly which varieties your market requires and that these varieties are well adapted to your area, your worry will be whether you can provide the essential attention at the times required. But in many areas the choice of varieties is not clear-cut and there is an element of experimentation involved. Then you should be more conservative in vineyard size and be willing to try only five to ten vines each of several varieties.

It is possible to go too far in experimenting. One or two vines each of many varieties may enable you to select certain kinds that do grow well for further planting. But orderly marketing of small batches of fruit is difficult, and quantities of fruit produced are not sufficient for the home winemaker to evaluate their wine potential. It is better to start with a small vineyard of ten to a hundred vines to which you can give proper attention than to plant a couple of acres without considering the costs, labor, and unexpected problems that may lead to a disappointing failure.

The prospect of a good harvest of grapes is exciting and potentially profitable. Remember that it does not just happen. Grape growers accept the fact that success requires regular chores, and that crop failure from frost, diseases, hail, or birds is a constant prospect.

Four Kinds of Berries

Strawberries, blueberries, blackberries, and raspberries offer a unique opportunity for a family of four to make a good supplemental income on 2 to 5 bearing acres. Berry crops require a lot of labor, but a family can usually manage all but the harvesting on a small acreage. Berry culture requires considerable knowledge of the crops and their care, and a large initial investment per acre. However, berries offer a high return per dollar invested because of a generally low supply and high demand in many regions of the United States.

The advent of direct farm-to-consumer marketing (pick-your-own) reduces harvest labor cost and provides the consumer with high-quality fruit. Future demand for berry crops promises to be high because of their appeal as sources of dietary enrichment and their varied uses.

Strawberries

The garden strawberry, *Fragaria x ananassa,* is a perennial herbaceous plant in the Rose family. The strawberry plant has a short thickened stem (called a crown), which has a growing point at the upper end of the crown and forms roots at the crown's base. The leaves are borne along the crown on long petioles (leafstalks), arranged in spiral fashion around the crown.

At the juncture of each petiole and the crown, a bud is borne that may grow into one of three structures according to the environmental stimuli the plant receives:

- **a runner** or specialized elongated stem, which normally forms daughter plantlets at every other node, if the temperatures are warm and the day-length long;
- **a branch crown** or new stem under the same temperature and day-length conditions under which runners are formed (some strawberries are runnerless forms and produce many or multiple branch crowns);
- **flower stalks** (scapes) bearing one to fifteen flowers, usually under short days and low temperatures ("June Bearers"). However, some strawberries flower under long days ("everbearers").

The strawberry fruit is the ripened receptacle bearing many small "seeds" (achenes). The seeds form following pollination of female parts of the flower, which are collectively arranged on the fleshy receptacle (modified stem).

Strawberry plant roots are usually confined to the upper 6 to 12 inches of soil. As such they are subject to soil moisture fluctuations and severe weed competition. Strawberry roots are vulnerable to nematode attack and several root-rotting fungi, notably red stele *(Phytophthora), Verticillium,* and *Rhizoctonia.* Frequently soil is chemically fumigated to kill these pests and many weed seeds before planting. Several strawberry cultivars are resistant to some of these soil fungi.

Strawberry plants grow on a wide variety of soils and tolerate wide ranges in soil acidity and composition. Generally the lighter soils (sandy) are chosen for early maturity. Strawberry soils should be reasonably high in organic matter and well drained. Strawberry roots will not tolerate poorly drained soils.

You must control weeds, mechanically or with chemicals, to successfully raise strawberries. One acre inch of water per week during the growing season is considered optimum, as is a slightly acid soil (with a pH of 5.5 to 6.0).

Strawberries grow in almost any climate, but it is critical to select cultivars adapted to your region and tolerant to its pests. Varieties developed at a certain latitude are generally adapted to within 3 to 5 degrees of the same latitude at the same elevation. A 1,000-foot increase in elevation reduces the mean temperature of 3 degrees Fahrenheit. A variety is adapted to a region if it grows and produces well in response to the region's prevailing environment and pests.

Climatic factors that influence strawberry adaptation are low winter temperatures (remember that strawberries are frequently covered by snow or straw in areas with cold winters), high summer temperatures, amount of winter chilling, and the length of growing season and days.

Major strawberry pests are viruses, nematodes, mites, fungi, and insects that attack the root, leaf, crown, or fruit. Consider how to control these—there are chemical sprays available, but use them judiciously. In many locations, you must water strawberries to protect flowers and green fruit against spring frosts.

Blueberries

Cultivated blueberries, *Vaccinium* species, are woody perennial bushes of the Heath family. The common varieties are tall multiple-caned (stemmed) plants with shallow and fibrous matted root systems. It takes three to four years for the plant to grow to mature size (you should prune them to a height of 5 to 10 feet, and 3 to 8 feet in width and depth). The single leaves grow in a spiral along new shoots, which grow from each stout cane. There may be several flushes of new growth each season.

The buds between each leaf and the stem become one of two types: leaf buds, which will expand into a leafy shoot in the following season; or flower buds, which will bear a cluster of two to twelve flowers the following spring. Each flower, if pollinated, will develop into a fleshy, many-seeded berry. A mature bush bears literally thousands of blueberry fruits.

Blueberries generally grow well in moist, but reasonably well-drained, acid soils with a high proportion of organic matter. You can modify mineral soils for blueberry culture by adding organic matter and soil acidifiers. Certain types of blueberries may be grown in acid mineral soils without soil amendments.

Blueberry roots are very fine, usually grow in the upper foot of soil, and are easily damaged by excessive fertilizer salts and standing water. Hence blueberry plants are frequently planted in low-lying, quite acid, sandy areas (the optimum pH is 4.5) on raised beds. Give special attention to weed and water control in blueberry fields.

Cultivated blueberries usually don't grow well where winter temperatures are lower than -20 Fahrenheit, where the growing season is less than 160 days, and where there are fewer than 1,000 hours of temperatures under 45 degrees during winter. Rabbiteye and southern species of hybrid blueberries do better in the warmer winters of the Gulf Coast and southern U.S.

Principal blueberry pests are the stunt and red ring spot viruses, two fungal stem cankers, a fungal root rot, a number of leaf and stem fungi, a bud mite, several fruit-rotting fungi and worms, and a number of chewing, sucking, and tying insects. Control these with resistant varieties and chemical pesticides.

Bramble Plants

Blackberries and raspberries (bramble plants) are woody, multiple-stemmed, usually thorny plants in the genus *Rubus* of the Rose family. Roots live for many years in brambles, but the stems usually live for only two seasons (biennial), vegetating in the first year and bearing fruit in the spring of the second season. Exceptions to this growth habit are certain raspberry varieties (everbearers), which fruit in the fall on first-year canes, and in the next spring on the same canes.

The canes may be either erect, semi-trailing, or trailing (the latter two need to be trellised), depending on the particular species of Rubus involved. The stout canes grow and branch, forming laterals during the first growing season. During the next season buds on the laterals grow and produce short shoots, which bear both leaves and flowers.

As in a strawberry, the female pistils surround a raised receptacle (or stem axis, called the torus in blackberries) in each flower. However, each pistil ovary develops into a fleshy drupelet bearing a single, hard, internal seed upon pollination. The drupelets adhere together, separating from the receptacle in raspberries, and including the central stem axis, when harvested, in blackberries.

Raspberries and blackberries will tolerate almost any soil type as long as it is well drained to a depth of 3 to 4 feet. Both crops grow well in soils having good organic matter content and water-holding capacity and a slightly acid soil reaction (pH 5.5 to 6.8). Soil grubs and certain weeds should be controlled before planting. Some bramble plants spread, so the row width must be controlled in subsequent years.

Raspberries are particularly hardy in areas with cold winters, withstanding temperatures of -35 degrees for red and -25 degrees for black (and purple) fruited types. However, raspberries are frequently injured in milder areas that have fluctuating winter temperatures.

Blackberries may be injured at -15 Fahrenheit, while thornless blackberries, boysenberries, and youngberries may be injured by 0-degree temperatures. Generally, raspberries are not too successful south of USDA minimum temperature zone 6.

Raspberries are especially sensitive to soil nematodes, viruses, and a number of insect and fungus troubles. Blackberries are troubled by the bacterial diseases crown and cane gall, the fungus diseases anthracnose, leaf and cane spot, orange rust, double blossom, and a considerable array of insects. Choose resistant varieties rather than attempting to baby the susceptible ones.

Berry Culture

To successfully grow strawberries, blueberries, and brambles, you need to understand some general principles. All of these crops grow on acid to alkaline soils (with a pH of 5.5 to above 7.0), except blueberries, which require an acid soil with a pH between 4.0 and 5.5. These crops grow best on well-drained soils that have a good supply of water. For top yields they should be irrigated in dry periods, and in strawberries spring-frost control by irrigation is desirable in most areas.

You can fertilize all of these with a complete fertilizer, such as 10-10-10, at the rate of 50 to 75 pounds of actual nitrogen per acre varied to the soil's fertility. This recommendation is for established mature plants. Be careful with fertilizers, however. You don't want to apply nitrogen to soil just before a rain, for instance, when most of it could wash away and into local water supplies. Consider organic alternatives, such as the use of compost.

Young and new transplants can be easily damaged by fertilizer. Apply lower rates initially and increase the rates up to the full amount on mature plants of blueberries and brambles. Rabbiteye blueberries probably should not be fertilized the first year.

Distribute fertilizer over the entire root zone just before times of greatest growth and fruiting. Prune the woody crops to remove dead and injured canes and to balance and distribute the fruit load. Most brambles fruit the second year on canes produced the previous season. Fall-fruiting red raspberries can be cut back in late winter and will fruit the following fall on new canes.

Training small fruits depends upon the plant's growth habit. Trailing blackberries and red raspberries are tied to a wire trellis for support; erect blackberries and black raspberries need not be. Erect blackberries are planted by covering pencil-sized root pieces with soil and permitting suckers to form a hedgerow, no wider than 24 inches. Red raspberries also sucker freely, and row width must be narrowed. Trailing and semi-erect types (such as the thornless varieties Smoothstem, Thornfree, Black Satin, and Dirksen Thornless) and black raspberries sucker very little and are propagated from cane tips that root in contact with the soil.

Planting density varies with the particular kind of berry. With black raspberries, blueberries, trailing and semi-erect blackberries, and strawberries grown in hill culture, the number of plants set is the same as the desired number to be fruited. Red raspberries, erect blackberries, and strawberries grown in the matted row are set with fewer than the desired number of plants and allowed to multiply.

Weeds, insects, and diseases must be controlled by a combination of cultural practices and pesticides. Soil fumigation before planting is expensive but pays well in controlling most weeds, soil insects, nematodes, and some disease organisms.

PLANTING AND PLANNING GUIDE FOR STRAWBERRIES, BLUEBERRIES, AND BRAMBLES						
Type	Planting distances		Planting stock	Years from planting to economic return	No. of bearing years	Average mature yield (lbs)
	Between rows (ft)	In the rows (ft)				
Strawberries						
Matted rows	3.5–4	1.5–2	1-yr runners (virus free)	1	2–3	1/row foot
Hill	4–5	0.5–1	1-yr runners (virus free)	1	2–3	1.5/plant
Blueberries						
Highbush	8–10	4–5	2-yr plants	3	25+	6–8/plant
Rabbiteye	10–12	6–8	2-yr plants	4	30+	12–15/plant
Brambles Blackberries						
Erect	8–12	2–6	root pieces	3	10–12	1.5/row foot
Trailing	8–12	6–8	rooted cane tips	2	5–10	9/plant
Raspberries						
Red	8–10	2–4	1-yr suckers (virus-free)	3	10–12	1.5/row foot
Black	8–10	3–4	rooted cane tips (virus-free)	3	3–4	1.5/plant

Fruits of these crops develop from flowers pollinated usually by local bee and insect populations. It is a sound practice to grow at least two rabbiteye blueberry cultivars in each ripening season for cross pollination. The other small fruits are self-fruitful, but can benefit from cross pollination by more uniform fruit maturation and increased berry size.

Using a straw mulch for winter protection is beneficial for strawberries in certain areas, and some growers use plastic mulch for moisture conservation and fruit protection. Blueberries and brambles also benefit from mulching, particularly in soils not optimally suited for these crops. Incorporating peat moss in the planting hole improves establishment and growth.

Economic Aspects

Berries can be expensive to produce. Looking at strawberries as an example, the Ohio State University Extension has estimated that 1 acre of strawberries you pick yourself costs more than $4,000 from planting to harvest, counting plants, mulch, equipment, pesticides, some hired labor, and fertilizer; that same acre of berries nets about $1,200. The University of Kentucky Cooperative Extension estimated that blueberries cost $7,682 per acre to plant a pick-your-own operation in 2008; that same acre would bring in a yearly income of $6,778 but not for seven years. Pennsylvania State University has estimated the per-acre costs of establishing an acre of strawberries as $6,387, also including the necessary plants, equipment, and labor. If that acre of berries produces the maximum of 10,000 pounds, and sells for $1.40 per pound, your net would be $7,613.

Strawberries begin to produce heavily in the year after planting, while brambles and blueberries start producing heavily in the third to fifth year. However, returns per acre and per dollar invested over several years can be the highest. Thus, you need to regard such enterprises as a long-term commitment. In planning for this commitment, there are a number of considerations.

LABOR REQUIREMENTS VARY FOR EACH MONTH OF THE YEAR. The jobs, number of hours, and time of year will vary with species, geographic region, plant density, and yield. If you are growing fruit on small plots, you might be able to do the work yourself. Alternatively, pick-your-own harvesting, where buyers harvest the fruit, provides an excellent means of selling your perishable crop quickly and receiving payment immediately. Pick-your-own harvesting is most advantageous within 25 to 50 miles of a large metropolitan area. Many rural and semi-rural areas attract city dwellers who own second homes or go on day trips.

VIRUS-FREE, DISEASE-RESISTANT, HARDY PLANTS ARE WORTH THE COST BECAUSE HEALTHY, VIGOROUS PLANTS PROVIDE THE HIGHEST YIELDS. Trellis materials should be of good and lasting quality. You should select chemicals based on local Extension recommendations. Take advantage of soil- and tissue-testing services provided by some state agencies.

EQUIPMENT PURCHASES CAN BE MOST CHALLENGING IN MATCHING PRICE AND SPECIFIC USES. For example, a tractor and sprayer are two indispensable implements. A small high-volume sprayer costs less than a low-volume sprayer. However, the high-volume sprayer will require more water and labor per application. The low-volume sprayer will require a larger tractor than the high-volume sprayer. Hence, if time is the limiting factor (as with weekend farming), you may need the more expensive equipment.

If you buy a tractor, be sure to match it to the equipment you will use with it. A reputable dealer can furnish parts and repairs in a short period of time.

Selecting a Crop

First, consider the amount of labor available within the family. Weather conditions and shorter days restrict working time in winter. Berry crops require a large amount of hand labor. If labor is limited, note which crops have high labor requirements during the same period of the year, and choose just one of them.

Next consider only a small acreage to start. This will reduce the financial risk as well as provide an opportunity to learn the biological, technological, and economic requirements before expanding. Plan ahead for expansion, rotation, and location of crops, particularly if you have a pick-your-own operation. Try growing just a row or two of blueberries or a patch or two of strawberries. See how they do on your property. See how you respond to the harvest times.

Unless you are planting just a few bushes or plants, you probably will have to borrow money to fund the initial cost of land, plants, and equipment. You should anticipate a rate of return greater than the interest rate you pay the bank. If you have limited cash and labor to invest, then a pick-your-own operation with strawberries and red or black raspberries would be a logical choice. If cash is not limited, choose strawberries, raspberries, and thornless blackberries. If ample cash and labor are available, you might grow strawberries, raspberries, thornless blackberries, and blueberries.

Since the species of berries are adapted to different regions, and varieties within a type of berry are adapted to smaller areas within a region, consult your local Extension service or agricultural experiment station for particular berry variety recommendations.

In land-use planning, set aside sufficient additional acreage for new plantings as present fruiting areas decline in vigor and productivity. The marketing season may be profitably extended in most areas if compatible vegetable, floral, nursery, and tree fruit crops are grown near the berry plantings.

Vegetables

Using all or part of your land to grow vegetables may be appealing regardless of whether you intend to feed your own family or to profit from it. Keep in mind that growing vegetables as a moneymaking venture is a serious business requiring many talents and skills. But the times could not be better for growing produce on a small-scale. Americans are clamoring for locally grown food. Not only do people want it at home; schools and colleges buy it in quantity as they try to ride this wave of locavores' demands. This is a great development for them—and for you.

Most vegetables yield a large return per unit of area, but growing them is a lot of work. Don't try them unless you have hours each day to tend to them, ideally with adequate labor available in the growing season. You should also be on hand to properly manage the planting.

Before deciding to grow vegetables, and which vegetables will thrive, determine whether your land will support them. You need a slightly acid, fertile, well-drained soil. Soil acidity and fertility can be altered by adding fertilizer and lime, but don't do this without having your soil tested. Wet soils can be drained with tiles, underground pipes made of clay, ceramic, or plastic that funnel water away from crops. Tile drainage goes back thousands of years, and it's a method farmer John Johnston brought to the United States from his native Scotland in 1835. Tile drainage has rendered much farmland useable in the Midwest, but tiles can filter fertilizers and pesticides to waterways and would require permits in many areas. Systems can be expensive and usually operate on larger farms. (A tile drain museum is open by appointment in Geneva, New York, where Johnston farmed.)

Without exception, vegetables grow best in full sunlight. Easy access to an abundant supply of high-quality water is also important or even essential in some areas. Expensive sprinkler systems aren't necessary unless the land is really dry, in which case you probably wouldn't be growing vegetables.

Finally, consider the length of the frost-free growing season and the temperature throughout the season. Growing zones have changed slightly as the climate has warmed. The USDA updated its Plant Hardiness Zone Map in 1990, and placed the map on the Internet in 2003. The zones range from 1 to 10, but within each, subzones alter the optimum planting dates slightly. See http://www.usna.usda.gov/Hardzone/hzm-ne1.html.

If you intend to make a profit growing vegetables, one of your first concerns should be how to sell them. You will probably choose to grow crops to sell fresh at a market, rather than for processing. This is because most of the processing crops are grown on large acreages so that they can be machine-harvested. A small-scale farmer using small machines or hand labor finds it very difficult to compete with these large mechanized farms. In contrast, vegetables for fresh consumption often can be grown by a small-scale farmer-grower or by a larger farmer using mostly hired laborers. The public wants fresh vegetables: they want sweet corn, tomatoes, peppers, summer squash (yellow and green), cucumbers, snap beans, lima beans, eggplant, acorn and butternut squash, okra, and more. They are interested in heirloom varieties of all of these. Also popular are the salad crops including cabbage (green, red, and savoy), lettuce (iceberg, leaf, Boston, Bibb, romaine, escarole, and endive), collards, kale, parsley, beets, and carrots.

Determining the Right Produce to Sell

You can set up a successful roadside stand if people like driving on your road and are willing to pay for higher-quality local produce. The most popular vegetables at roadside sales are sweet corn, tomatoes, and salad crops. An important factor in their popularity is that they are so highly perishable. Roadside marketing makes it possible to reduce the time interval between harvesting and consumption, thus making it possible for the consumer to have the highest possible quality.

Your crop mix for roadside marketing can also include peppers, cucumbers, yellow and green squash, new potatoes, cabbage, muskmelons, watermelons, beets, carrots, and cauliflower. Strawberries also fit into roadside marketing, as do pumpkins grown for the Halloween trade, Indian corn, and some winter squash.

When developing a pick-your-own operation, give thought to what vegetables are best suited to this system. Since many inexperienced people will be harvesting, crops that are easily harvested and do not require difficult decisions on their being at the proper stage for harvesting are best adapted to a pick-your-own operation. Tomatoes, snap beans, and cucumbers are examples.

On the other hand, some experience is required to be able to determine when sweet corn is at the proper stage of maturity for harvest. Customers without sufficient knowledge not only end up with lesser-quality produce, but can also cause significant damage to a planting.

Once you have decided to grow vegetables and have determined a marketing plan, consider what crops will grow well in your area, the potential demand, the likely monetary return from each, and what specific varieties are best adapted to your area.

If you plan to grow several vegetables, a comprehensive plan for using the available space and the planting sequence should be developed before putting in the first seed. Also, decide what might be grown in future years so a crop rotation system can be developed.

If space is limited, you will probably want to grow crops that will give the largest return per unit of area. Examples are tomatoes, peppers, summer squash, and cucumbers. On the other hand, sweet corn, pumpkins, and winter squash need a great deal of space and may not provide as healthy a return on your investment.

Unending Harvest

Sound planning before the time to plant can result in a continuous harvest throughout the season. One way to use space effectively is to make a sequential planting of crops with different temperature requirements and/or different rates of maturing.

For example, an early crop of lettuce, cabbage, or other cool-season vegetables can be followed by snap beans or summer squash. These in turn can be followed by another planting of the cool-season crops, such as cabbage and lettuce.

Continuous harvests of a single crop can be achieved by successive plantings of the same variety at a regular interval, such as every two weeks. The same result may be achieved by simultaneously planting several varieties of a crop that are known to mature at different rates.

Use cultural practices that will let you minimize the amount of labor required. These might include the use of various mulches, herbicides, or drip irrigation. Mulches can significantly lessen weed problems and reduce the need for other weed control methods. They also can result in warmer soil, thus earlier crop maturity, conservation of moisture, and reduced disease.

SUGGESTED PLANT SPACING, NUMBER OF SEEDS OR PLANTS REQUIRED, AND AVERAGE YIELD OF THE COMMON VEGETABLES

Vegetable	Spacing (inches) rows	Plants	Plants or seed per 100 feet	Average yield expected 100 feet
Asparagus	36–48	18	66 plants or 1 oz.	30 lb.
Beans, snap bush	24–36	3–4	½ lb.	120 lb.
Beans, snap pole	36–48	4–6	½ lb.	150 lb.
Beans, lima bush	30–36	3–4	½ lb.	25 lb. shelled
Beans, lima pole	36–48	12–18	¼ lb.	50 lb. shelled
Beets	15–24	2	1 oz.	150 lb.
Broccoli	24–36	14–24	50–60 plants or ¼ oz.	100 lb.
Brussels sprouts	24–36	14–24	50–60 plants or ¼ oz.	75 lb.
Cabbage	24–36	14–24	50–60 plants or ¼ oz.	150 lb.
Cabbage, Chinese	18–30	8–12	60–70 plants or ¼ oz.	80 heads
Carrots	15–24	2	½ oz.	100 lb.
Cauliflower	24–36	14–24	50–60 plants or ¼ oz.	100 lb.
Celery	30–36	6	200 plants	180 stalks
Collards & Kale	18–36	8–16	¼ oz.	100 lb.
Corn, sweet	24–36	12–18	3–4 oz.	10 doz.
Cucumbers	48–72	24–48	½ oz.	120 lb.
Eggplant	24–36	18–24	50 plants or ⅛ oz.	100 lb.
Kohlrabi	15–24	4–6	½ oz.	75 lb.
Lettuce, head	18–24	6–10	¼ oz.	100 heads
Lettuce, leaf	15–18	2–3	¼ oz.	50 lb.
Muskmelon (Cantaloupe)	60–96	24–36	50 plants or ½ oz.	100 fruits
Okra	36–42	12–24	2 oz.	100 lb.
Onions	15–24	3–4	400–600 sets or 1 oz.	100 lb.
Parsley	15–24	6–8	¼ oz.	30 lb.
Parsnips	18–30	3–4	½ oz.	100 lb.
Peas, English	18–36	1	1 lb.	20 lb.
Peas, southern	24–36	4–6	½ lb.	40 lb.
Peppers	24–36	18–24	50 plants or ⅛ oz.	60 lb.
Potatoes, Irish	30–36	10–15	6–10 lb. of seed tubers	100 lb.
Potatoes, sweet	36–48	12–16	75–100 plants	100 lb.
Pumpkins	60–96	36–48	½ oz.	100 lb.
Radishes	14–24	1	1 oz.	100 bunches
Spinach	14–24	3–4	1 oz.	40–50 lb.
Squash, summer	36–60	18–36	1 oz.	150 lb.
Squash, winter	60–96	24–48	½ oz.	100 lb.
Tomatoes	24–48	18–36	50 plants or ⅛ oz.	100 lb.
Turnip greens	14–24	2–3	½ oz.	50–100 lb.
Turnip roots	14–24	2–3	½ oz.	50–100 lb.
Watermelon	72–96	36–72	1 oz.	40 fruits

Organic materials such as straw and leaves or specially manufactured polyethylene or paper products can be used as mulches.

Herbicides, or chemical weed killers, may be valuable to a small-scale operation. However, no one herbicide is effective or legally approved for use on all vegetables. Be careful to use only those chemicals approved for weed control for a given crop.

Using a material not labeled for a crop is unlawful. This can damage the crop and pose a possible problem to the people consuming it. Strict adherence to instructions on the container label is essential.

Drip or trickle irrigation is a good watering technique. It involves the use of a plastic pipe or hose laid very near a row of plants. The pipe or hose has minute holes that allow water to gradually seep out and wet the area immediately around the plants. This provides a continuous supply of water to the crop and delivers the water to where it is of greatest use. Drip irrigation also greatly reduces the amount of water required to grow a crop—which is particularly valuable in arid areas.

The drip system is especially effective when combined with use of paper or polyethylene mulch. The irrigation hose is centered under the mulch along the plant row. Thus the mulch reduces water loss due to evaporation from the soil surface.

Plant Pests

Keep a sharp eye out for insects and diseases or you may lose all or part of the crop in a short time, or suffer from a poorer-quality crop. Seek out information about pests and how to control them from the USDA and state agricultural experiment station brochures, county Extension agents, and agricultural supply agents. Many growers will be happy to talk with you about their experiences and can be a tremendous help.

The equipment you need to grow vegetables on an acre or so depends in large part on the size of operation you envision. Basically you need some means of preparing the land for planting, of tilling or cultivating during the season, and of applying fertilizer and controlling pests. It may also be desirable to have mechanical equipment available for seeding and/or transplanting. A wide variety of equipment is available for all sizes of farming operations.

Modern farm equipment is extremely expensive, and should not be purchased if you can hire the job done for less. For a small acreage you may find it advantageous to hire another farmer to do your plowing and soil tillage and then buy a new or used small farm tractor for the row crop work of cultivating and spraying.

Most small tractors designed for mowing lawns are not powerful enough to plow satisfactorily, and do not have enough clearance to permit driving over a partially or fully grown crop row for spraying, cultivating, and the like. You will need at least a 30-horsepower tractor to plow 9 inches deep to turn over the soil properly and make a good seedbed.

A small heated and ventilated greenhouse is essential if you want to grow transplants for earlier planting. You can advance your harvest season by several weeks if you raise your own transplants of tomatoes, peppers, melons, cabbage, and lettuce. It takes about six to eight weeks to grow transplants of peppers, tomatoes, cabbage, and lettuce. Melon plants are ready for field planting in three weeks. An inexpensive greenhouse can be built by covering a pipe frame with polyethylene.

The foregoing gives general information on requirements for growing vegetables for your gratification or for developing a small venture for profit, either on your own or in combination with neighboring farmers.

Growing Organically

If you can tell your customers that you didn't use any chemicals on your fresh produce, you have a leg up. Growing organically is not difficult on a small scale. For some plants like squash you must be vigilant and time planting to avoid borers that kill. For others like tomatoes you might be out with your pepper-based spray to keep Japanese beetles off. It's worth a try.

Organic gardening is founded on the principle that organic matter—that is, composted plants, leaves, vegetable scraps, ground-up sticks, etc.—improves the soil, acting as both fertilizer and conditioner. It makes plowing and cultivating easier. It also increases the nutrient reserve and water-holding capacity of sandy or clay-type soils.

The gardener derives several benefits by mulching with organic matter. It reduces erosion caused by runoff of rain or irrigation, increases infiltration of water into the soil, and conserves this moisture by reducing evaporation. Organic matter helps to suppress weed growth.

Some good organic materials to use as mulch include leaves, lawn clippings, fresh sawdust, fine wood shavings, pine needles, chopped straw, ground corncobs, shredded tobacco or sugarcane stems, peanut hulls, and cottonseed hulls. These materials do not add important amounts of nutrients or have a significant effect on the pH (relative acidity) of the soil.

The dead vegetable plants in your garden may be chopped down and left on the ground as a protective mulch during winter. This trash mulch reduces erosion and improves organic matter content of the soil when the garden is prepared for planting in spring.

All year, use kitchen scraps and yard cuttings in compost piles. Enclose the piles in nailed-together pallets or scrap lumber. Cooperative Extension offices or garden stores sell recycled plastic compost bins with tight lids for about $50. These keep out animals and provide enough heat to accelerate the breakdown process. You ideally want an aerobic compost pile. But even in the dead of winter when you forget to stir it up, anaerobic breakdown works, too. It just takes longer.

Unmulched areas in gardens and fields, not occupied by growing crops, should be planted to green-manure crops such as rye, ryegrass, millet, sorghum, or crimson clover. They will reduce leaching of nutrients and increase organic matter for the next crop as they are worked into the soil.

Organic waste materials such as leaves, manure from livestock and poultry, treated sewage sludge, and the organic portion of urban trash collections can be used as fertilizer, mulch, or compost. Some cities accumulate leaves in huge piles during fall collection periods. After several months of composting, the material is available at little or no cost to gardeners. This is a practical way to reduce environmental pollution and supply organic material for gardens and farms.

Timing to Avoid Insects and Disease

Time of planting is important in avoiding losses by diseases and pests in certain regions. Since seed corn maggots destroy early plantings of beans and corn, you should delay planting until the soil warms. Early-maturing varieties of sweet corn can avoid the worst earworm problem. Delay plantings of summer squash to avoid early-season activity and resultant damage by the squash vine borer.

During recent years, plant breeders have made tremendous contributions to agriculture by developing new varieties resistant to diseases. These allow large yields of high-quality crops to be produced without the use of chemical sprays.

When planning for vegetable production in a home garden or commercial enterprise, consult your county Extension agent or seed catalog for information on disease-resistant varieties. Excellent new introductions are available each year. Some vegetable crops are highly subject to damage by pests or disease organisms. Others are relatively pest-free. The beginner should first plant only trouble-free crops, trying the more difficult ones after gaining some experience.

Attacks by cutworms can be prevented by placing a simple collar of stiff paper (cut from a drinking cup or milk carton) around newly set tomato, cabbage, and pepper plants—and even sweet corn. The collar should extend about 1 inch into the soil and 2 inches aboveground.

Slugs that emerge at night from hiding places in wall crevices, loose mulch, piles of plant stakes, or trash can be trapped under pieces of board, shingles, or flat stones laid in the garden. Lift them each day and destroy the slugs.

Slugs are attracted to shallow vessels partially filled with beer into which they crawl and expire. Slug baits moistened with a teaspoon of beer will be twice as effective.

An aluminum foil mulch around low-growing plants reflects the ultraviolet rays from the sky and repels flying insects (including aphids, leafhoppers, thrips, Mexican bean beetles, and cucumber beetles) from landing on the plants. This method protects summer squash, Chinese cabbage, lettuce, and peppers from virus infection transmitted by aphid feeding, and beans and cucurbits (from the Gourd family) from chewing and sucking insects.

Black polyethylene mulches, used extensively by commercial fruit and vegetable growers, help to control weeds, conserve moisture, and prevent leaching of fertility in the garden. They also keep the produce from resting on the soil, thus reducing rot infection from soil contact.

Blacklight traps are frequently advertised for control of insect pests in gardens and on farms. Although great numbers of moths and other insects are attracted to individual black lights and captured in the attached traps or killed on electric grids, there is little or no reduction of the pest insects that attack vegetables.

Sometimes insect pests in the vicinity of the trap will be greater than normal. Insects attracted to the light may not enter the trap, but instead may linger to lay their eggs in the vicinity. Likewise, certain bait traps—as for the Japanese beetle—may increase the infestation in the trap's vicinity.

Routine inspection and handpicking of tomato hornworms on a small planting of a dozen or so tomato plants is highly effective and less time consuming than preparing and applying a spray. In some years, hornworms may not appear at all. Handpicking can also eliminate small infestations of squash bugs, Mexican bean beetles, and potato beetles.

How to Market Produce

Despite the mass distribution system in which we live, there are several ways for a backyard gardener to sell surplus produce. Bypassing whole-sale marketing channels does not ensure a better price, but it might be the only way you can market your goods.

The growth of farmers' markets and community supported agriculture (subscription farming) has made marketing fresh vegetables easier than in the days of setting up a quiet roadside stand. You can make your vegetable business large or small, depending on your willingness to go to markets and reach out to subscribers and how much time and help you have.

Of course, roadside stands remain a nice way to save fuel and time if your farm isn't far off the main roads. You might want to run a pick-your-own farm, which are very popular on weekends as long as you don't mind the crowds. You may deliver to local stores. You may also choose to sell through a broker-shipper who buys your produce, but it's easier today to sell directly to consumers instead of paying a middleman. Also, people like meeting farmers.

Roadside Marketing

Selling produce and related products at the roadside lets you sell at near-retail prices. The venture is relatively easy to start with limited capital. As with any type of business, roadside marketing does have its drawbacks. Some potential entrants may not have the personality to deal successfully with people or possess the managerial skill to operate a business; others may not be prepared to put in the long hours often required for six or seven days a week. The individual operator must also be prepared to bear the cost of promotion, merchandising, quality control, and customer service.

Your enterprise could range from placing a few tomatoes on a card table under a shade tree to setting up a permanent facility along the lines of a supermarket. Experienced roadside marketers would recommend that the beginner start off in a small way.

Location appears to be a key factor. Before you start up a market, determine the site's accessibility to potential consumers. Ideally, roadside markets should be near towns or cities, on heavily traveled highways. Many operators recommend locating on the inbound side of a road to a city, or on a straight stretch offering good visibility. Attention to ease of entry, parking, and egress is important. Neatness and attractiveness of the site and structures is also a plus.

Regulations

State and local zoning regulations, licensing requirements, sales taxes, and labeling, sanitary, weight, and Sunday operating policies vary. Check these requirements with the proper authorities. For example, construction of buildings near rights-of-way may be restricted and access permits required.

Because customers prefer to make all their purchases at one stop, it is important to carry several items. Top sellers among vegetables are tomatoes, sweet corn, green beans, and melons. Any fruits—particularly apples, peaches, and strawberries—sell well and should be handled in season if possible. More commercial operations may include eggs, flowers, honey, Christmas trees, and other specialties. In any event, customers expect to find locally produced items available in season. This calls for scheduling planting and harvesting to ensure adequate supplies for sale.

High-quality, fresh, homegrown fruits and vegetables are the backbone of a roadside market. This implies careful grading and sorting procedures, coupled with measures to maintain condition of the products on display. However, customers often differ in their requirements for quality as well as their ability to pay. Market operators often find it good practice to offer two or more grades of the same commodity. Thus, both quality shoppers and price shoppers can be satisfied. In any case, represent products honestly to foster repeat sales. This is especially important with prepackaged items.

Preservation of quality while on display may entail refrigerating certain commodities. Highly perishable items like sweet corn or strawberries should be cooled. Others, particularly leafy greens, should be sprinkled periodically to compensate for moisture loss. The means for maintaining condition—icing or mechanical refrigeration—will depend on sales volume of the operation. A small volume, of course, will not justify a large investment in mechanical equipment and facilities.

Pricing policy among roadside stand operators varies, but prices charged tend to fall between those in nearby city retail stores and wholesale markets. Selling above the retail price may be justified where quality is exceptionally superior. However, customers expect a somewhat lower price to compensate for the transportation and delivery services they assume.

Advertising

Most roadside marketers stress the importance of advertising to attract customers. The extent depends on the type of market, location, commodities and volume handled, and other factors. The best advertising is "word of mouth" by satisfied customers; however, since it is based on performance, it takes time to establish. Signs placed on roads leading to the market are often effective on well-traveled routes. Newspapers and radio are frequently used by operators to good effect. Direct mailings and handouts are other methods.

There are many other facets to the organization and operation of these markets. For example, many operators have found a pick-your-own operation a good adjunct to a roadside stand business. Others, especially those with small volume and few items to offer, have experienced satisfactory results using nonattended self-service stands. Honor selling, while not without hazards, can under some circumstances be profitable and provide a start to a full-fledged marketing operation.

Fortunately, much help is available to the prospective roadside marketer. Many state and county Extension services and universities provide information through meetings and publications. Some state departments of agriculture are also active in this area. Experienced roadside operators can provide some helpful hints and guidelines.

Roadside marketers in a number of states have organized into associations to provide themselves with information and services they could not supply on their own. Major services of the cooperative groups are advertising, promotion, centralized purchasing of marketing supplies, and setting industry standards. State departments of agriculture often assist by providing inspection and certification services to the group. Members who qualify can display trademarks or certificates attesting to their compliance with the association's code of ethics.

Pick-Your-Own

The pick-your-own method of selling, as the name implies, is where the customer does his own harvesting in the field or orchard. This concept of consumers bypassing the usual marketing system has been growing in popularity. From the customers' viewpoint, pick-your-own offers a chance to select the desired size and quality of fruits and vegetables. Assured freshness of product is, of course, a major motivation for consumers to do their own harvesting. Expectation of reduced prices is another.

From the grower's standpoint, the pick-your-own method greatly reduces the need for harvest labor. In addition, grading, packing, and storing functions are eliminated. Thus, net returns may be increased. Consumer harvesting is especially adapted to crops that tend to mature and be harvested all at once, or where ripeness may readily be determined by color or size. Fruits and vegetables that are commonly canned or frozen in homes are good candidates.

Pick-your-own crops include apples, peaches, grapes, cherries, cranberries, strawberries, peas, green beans, potatoes, and even Christmas trees. Items whose stage of maturity is more difficult to judge, such as melons and sweet corn, appear less suitable for consumer harvesting.

Location and distance from population centers may be less critical to a pick-your-own operation than for other direct marketing approaches. Location must necessarily be tied to the land where the crop is grown. Experience indicates that customers are more likely to travel longer distances to harvest their favorite crops and often look on the activity as a recreational outing or a way to savor their rural heritage.

A chief concern of pick-your-own growers is the possibility of damage to plants or trees by inexperienced handlers. However, it has been shown that with proper planning, excessive damage to the crop can be minimized. Most customers, given a little instruction, will treat plants or trees with care, and reductions in yields—if any—should be minimal. Many experienced pick-your-own operators report that customers have a tendency to pick all the fruit or vegetables before them even though the items of better quality may have been spot-picked beforehand. Apparently, the satisfaction gained from picking one's own produce often overrides grade or quality considerations.

Another concern is the risk of financial loss or costly litigation from personal injury to customers while on the property. Although the possibility of accidents may seem small, the consequences of such occurrences can be devastating. It is essential to make your property as safe as possible. Preventive steps might include prepicking fruit from tall trees prior to public entry, keeping ladders and other equipment in good repair, fencing off ponds, ditches, and other possible hazards, supervising children—perhaps in a designated play area—and posting ground rules promoting safety.

Accident and liability insurance coverage is a must in any pick-your-own operation. Consult legal and insurance advisers before starting operations.

Attracting Customers

Potential customers need to be informed when crops are available for harvest and the times when picking will be permitted. Providing directions to the property is also vital. A number of methods for presenting and publicizing the operation, including ads in daily newspapers and local radio commercials, might be considered. However, be careful to tailor the coverage to the volume of production available and the capacity to handle the anticipated clientele; otherwise, chaos can result.

For established small operations, you might develop a mailing list of prior customers. Mailed announcements of crops and picking dates may be sufficient to move the available volume. Road signs are also useful to announce that picking is in progress, as well as to direct customers to the site. On well-traveled highways, signs alone may be sufficient to attract the desired number of customers. Placement of signs must conform to state and local regulations.

Parking for customers' cars deserves special attention. Provide safe and convenient space for the peak volume of cars anticipated. Weekends are usually the busiest periods. Keep in mind that pick-your-own operations require a longer stay by customers compared with other retail selling methods. Plan the relationship of the parking area to picking sites with a view to reducing the carrying distance of often bulky and heavy produce. Locating the checkout point close to the parking area makes security and theft prevention easier.

Prices, of course, need to more than meet production costs yet be competitive for the area. Customers should sense that the operation is sharing the picking-cost savings with them. One basis for establishing prices might be to use retail store prices less harvesting and delivery costs, plus the additional marketing costs incurred under the pick-your-own operations. You might offer discounts for volume purchases. Sell by weight, volume, or count. Many operators find selling by weight a fair basis for pricing most items. Often customers wish to use their own containers to save money.

To learn more, contact your Cooperative Extension, state agriculture department, or land-grant college.

Farmers' Markets

Farmers' markets have exploded in popularity during the last decade or so. They provide a meeting point for urban and rural populations. They keep growing in popularity because consumers want fresh products directly from growers. The number of farmers' markets in the United States grew by 79 percent between 1994 and 2002. Today there are at least 3,100 farmers' markets where about 19,000 farmers sell produce, the USDA reports.

For information about farmers' markets, call the USDA's Farmers Market Hotline at 1-(800)-384-8704 or visit www.ams.usda.gov/farmersmarkets.

The USDA has started a program for farmers to set up markets on federal land. It also offers farmers' market coupons to low-income mothers and children through its Women, Infants and Children program, and to senior citizens, so that they may buy fresh produce at markets.

Municipalities often set up farmers' markets on certain days of the week. The markets are a place to sell your fresh produce and other products such as maple syrup, honey, eggs, flowers, crafts, and plants.

Most markets charge a rental or service fee, usually quite moderate. Variations may exist because of the number of stalls or vehicles used, location within the market, seniority of the vendor, volume of products sold, and whether the market is used on a seasonal or daily basis.

Most markets operate seasonally. Some may open year-round. The typical farmers' market operates one or two days a week. Fridays and Saturdays are usually big market days.

In Seattle, the Neighborhood Farmers Market Alliance runs four popular markets during the warm months. More than 140 growers from all around Washington State go into the city four days a week at four different locations.

In Camden, Maine, you can sell produce at twice-weekly farmers' markets during the spring and summer, and once a week at an indoor market in the winter. In California, more than 350 cities and towns operate farmers' markets, under the guidance of the state agricultural department. You could sell most days of the week in that state. For instance, if you live in the San Joaquin Valley, you might be selling on Sunday and Tuesday in the Fresno River Park, on Thursday in Modesto, and on Saturday at one of three locations in Bakersfield—and there are still a few dozen other markets to choose from. In Indiana, there are fifty-seven farmers' markets around the state. They are located in parking lots, malls, plazas, and at intersections in cities and towns.

Before venturing into a farmers' market operation, visit the market ahead of time. Since many markets open only one or two days a week, affiliation with several markets may be necessary to gain more sales days. Estimate your costs, including the fuel consumed to drive in from your property.

CSA, or Community Supported Agriculture

Another booming method of selling your produce is the CSA, which stands for "community supported agriculture," a direct-to-consumer marketing arrangement that permits household consumers to purchase advance shares of a farm's production in return for regular (usually weekly) deliveries during the growing season. CSA operations have experienced a dramatic rise in popularity in the United States during the past several years, expanding from an estimated 60 operations in 1990 to approximately 1,100 operations by 2006.

CSAs are not new. In the late 1960s a number of people who wished to avoid the traditional marketing channels, and who desired less expensive produce, joined to form food-buying clubs. These clubs often bought directly from farmers. Thus they were not faced with grading regulations, and they could buy more fully field-ripened produce for their members. Many clubs were loose neighborhood organizations. They called farmers and asked for the names of other farmers who might be interested in direct selling. In some cases, most of the purchasing took place at a local farmers' market. In many of the groups, members had to order ahead of time.

CSA Resources for Farmers

The Alternative Farming Systems Information Center, part of the USDA National Agricultural Library, offers many resources for farmers to start and maintain a CSA (community supported agriculture). Visit http://www .nal.usda.gov/afsic/pubs/csa/csa.shtml.

See also these helpful resources:

Center for Agroecology and Sustainable Food Systems at the University of California at Santa Cruz: http://casfs.ucsc.edu/education/instruction/ tdm/contents.html

College of Agriculture and Life Sciences, University of Wisconsin-Madison: http://www.cias.wisc.edu/economics/community-supported-agriculture-farmsmanagement-and-income

How to Set Up a Vegetable Box Scheme. Briefing Paper. Soil Association, 2007. See http://www.soilassociation.org/web/sa/saweb.nsf

Tegtmeier, Erin, and Michael Duffy's publication on CSAs from the Leopold Center for Sustainable Agriculture: http://www.leopold.iastate .edu/pubs/staff/files/csa_0105.pdf

Wilson College Center for Sustainable Living offers a publication for $10. For information see http://www.wilson.edu/wilson/asp/content .asp?id=1275.

Sale to Retail Stores

Most direct sales to retail stores are of seasonal crops, many of which have high handling and shipping costs. Perishability or weight may be major factors: examples are flowers, strawberries, cantaloupes, peaches, and pumpkins. The grower will incur a greater portion of the handling and shipping costs.

Selling directly to stores cuts handling and warehouse costs, supplies better-quality produce, and supports local production. An advantage to you, the grower, would be that you would develop a stable season-long outlet. You can reuse your containers after you make the delivery. An individual grading system would allow you to charge premium prices. If the store displays your name on the shelf with your produce, this advertising may lead to increased farm or outlet sales.

There are some problems, however. The initial contract for store delivery may be hard to make if the store does not have experience with you. You may have to hire an additional worker to do the delivery. Under some arrangements the retailer may give the grower only a short lead time before stating how much produce he wants, and when he wants it. The arrangement may demand precise timing of the picking or extend the harvest scheduling.

Bad weather may slow your production. Problems also arise because your fruits and vegetables are traveling to the market at a riper stage than those that go through other marketing channels.

Growers may tend to underestimate costs and set the price too low. To avoid this, estimate how much it will cost to pack and deliver the harvest. If you harvest something small and delicate, like raspberries, consider the cost of using small packing units as well as the extra work to fill them carefully.

Consider the return on your time investment while negotiating the details of the contract, handling and delivery, and mileage. Compare those costs to marketing produce through other outlets. Don't disregard such minor points as the time of day you deliver, as well as the possibility of selling numerous products at the same time.

Most supermarkets do not make cash payment. The managers send a delivery ticket to headquarters, where staff members match it against your invoice form. Then they mail the check. Schedule a conference before the season starts to make sure both parties understand the contractual agreement. A postseason conference is a good idea. This maintains smoother working relationships. Some supermarket chains have field buying offices that may help facilitate the process.

Organic Products

New laws defining organic foods require that your organic produce be certified as organic. This may be done by personal inspection by the retailer or through certification of the grower by a regional organic farmers' association. Check with your region's Cooperative Extension or a farmers' group for help.

Organic gardening is a way of life that extends far beyond the marketplace. You may agree with those growers who seek a way to distribute produce and not to focus on personal gain. Their goal is a personal reward of enriched soil and good food rather than mere dollars.

The sale price of fresh organic products varies greatly. The retail price may be higher than nonorganic produce because of the small scale of many retail outlets. Commodities that cost no more to grow organically than with petrochemical inputs may not, in some cases, bring a higher price in specialty outlets.

What is being sold often is taste and a belief. In some cases, this may conflict with consumers who expect to see perfect-looking produce all the time. In other cases, buyers may become skeptical if they think the prices of organic produce are too high.

Marketing of organic products is competitive. Individuals interested in such marketing should investigate the ramifications. A starting point could be to contact your regional organic farmers' association.

Local Restaurants

If you sell to a restaurant, you will make arrangements one on one, on a personal basis. High-quality produce is a must. Due to variation in the menu, not all commodities may be sold at the same location every day.

Be prompt. If the food is on the menu, the restaurant owner must buy it somewhere. If the grower neglects to deliver, the restaurateur has to take the time to go find the food elsewhere. Volume of delivery also may vary greatly between days of the week. Remember that you are competing with the frozen and canned food industry. Not all restaurant patrons or restaurant owners are discriminating about the produce from different marketing channels.

In relation to the volume sold, consider mileage driven, number of stops, length of time at each stop, and the need for dealing on a personal basis with the restaurant employees or owner at each stop. At times, you might arrange to trade the produce for restaurant meals.

Maintain constant contact with the restaurant owner or purchaser for a small institution to ensure future sales. Due to the loose nature of the arrangement, you may face the problem of competition from other growers, some of whom may be relatives or close friends of the restaurant owner. Selling to larger institutions and restaurants may be on a much more formal basis and call for a written contract. Many state-run institutions have their own source of supply.

Canning Foods

Some families take advantage of marketing surplus home produce by canning vegetables and fruits as specialty items. You can sell these specialties at roadside stands, at the farmers' market, through local novelty stores, and at restaurant gift shops.

It's possible to form an association with other producers and market your products under one label. Can home specialty items in small-volume glass jars or cans. Some customers buy the glass jars of fruits or vegetables for decorative purposes. When canning, exercise caution to prevent contamination. Comply with local, state, or federal laws.

If you run a mail-order business, remember that shipping and packaging costs may add substantially to your production costs. Consider also the cost of advertising. You might want to join a direct-mail association.

All of these costs could put the price of the final product way above the supermarket price. Remember that regardless of the condition of the produce when shipped or mailed, your buyers judge what they see when it arrives.

Community canning centers, which were quite common between the Depression and the early 1950s in some areas, enabled a number of neighbors to operate a facility to can surplus vegetables or fruit. These centers are rare in the early twenty-first century. Besides the usual fruits and vegetables, it is also possible to can fish, poultry, and red meats. Due to contamination problems, however, professional supervision and assistance may be necessary.

Auction Markets

Of all commercial marketing outlets, the produce or fruit auction probably offers the best opportunity to the small grower. Most auctions are open to the public. Quantities may be small or large. You can sell your produce to buyers at an auction, and then you can buy produce you don't grow to supplement your own roadside stand.

The typical auction market has facilities that permit one or more lines of vehicles such as pickup trucks or cars to pass through the "auction block" at the same time. For large producers, the fruit or produce may be a sample of that in the field or on a tractor trailer nearby. Buyers can check the quality of the fruit or produce as the handler opens the containers for their observation. The buyers look for quality, so small growers can compete for price without feeling discriminated against.

Chain stores, operators of small locally owned stores, and private individuals are often among the buyers. Revenue to operate the auction comes from a "charge per unit" or a percentage of the sales price.

Marketing Co-ops

Small farmers can join forces to market their produce by incorporating as a marketing cooperative. Members are limited to individual producers, whether they garden in the backyard or farm many acres. By forming co-ops, these members are attempting to attain the highest prices for their produce.

Services by the co-ops range from simple pooling to operations that include many of the following: harvesting, grading, packing, cooling, storage, transporting, and providing marketing and production supplies. Some cooperatives also are engaged in processing.

Members bring their produce to a central location, where the co-op usually commingles products, thus reducing each member's risk of loss due to spoilage, market decline, and so forth. Some of these co-ops link to larger regional co-ops in the hope of establishing a reputation for high-quality products. To this end, many have their own brand names that they market regionally or nationwide.

As a member, an individual has voting privileges in establishing operating procedures and also via capital stock ownership of the co-op. The co-op returns any excess profits (after maintaining a reserve fund) to members in the form of patronage refunds based upon the individual's participation.

Ornamental Plants

It's summer, and everyone wants to buy begonias and impatiens for their yards. Or it's Mother's Day and people are ordering Hawaiian leis like crazy. Or it's autumn, and people are pounding the pavement for potted chrysanthemums. Small growers can take advantage of this seasonal enthusiasm that grips Americans, providing the vast consumers' market with such hardy plants or flowers, in unique varieties that are adapted to specific climates. If you are willing to work hard and to forgo a financial return for at least a year or two, consider selling your stock to local nurseries or opening your own nursery. It is a nice way to run a business.

But don't expect windfall harvests of either money or plants. The main pleasure is not financial. Instead, it comes from the opportunity to work from a few to many hours a day with living green things, toward the specific goal of attracting a following of customers by providing the plants they seek—as well as occasional specialties they will be excited to discover. There can be a modest return, enough to help offset the cost of a college education or to provide a retirement cushion.

You can grow plants in rural, semi-rural, or suburban locations; the only requirement is a modest greenhouse with a source of heat.

Colorful Annuals

The raising and sale of colorful annuals has much to recommend it because the grower's season is fairly short, starting in the North in late February and ending about July 1. During this period the grower cannot be away from home overnight unless someone is on hand to maintain the heat and tend the plants. Because the spare-time grower works on a small volume and margin, it is usually not feasible to hire outside help. Even if it were, some growers report, the bookkeeping involved may rule against it.

Standard annuals, the ones customers know best, are usually dependable sellers provided they are in bloom at selling time. These include marigolds, petunias, salvia, and zinnias. Begonias and impatiens, though more costly to raise, always seem the first to sell out.

There is a smaller market for hardy perennials, but it should not be ignored as an adjunct to sales, particularly if the plants are brought into bloom during the peak selling season. These include columbine, bleeding-heart, delphinium, certain phloxes, and bearded iris.

An important consideration is that hardy perennials can be grown exclusively in cold frames and do not require a heated greenhouse, though the latter can spur early-spring growth or serve as a propagation aid. You can grow hardy perennials on a small scale, selling them to garden centers. This requires only cold frames, about an acre of land, and few overhead costs.

Corner a Market

It is essential to raise plants that customers want. These might not be your personal favorites. Start on a small scale, first getting to know both plants and their market potential, and reinvesting all profits in the business during a building-up period. Growing too many kinds of plants can be disastrous, the only beneficiary being the grower's own garden. It is better to sell out of a plant quickly than to have many leftovers, although the grower can sometimes dispose of leftovers toward the end of the season by contracting to plant other people's gardens.

Concentrate on plants that the big wholesalers have overlooked or disdained because they require special care, but for which there is a market. Gardeners are interested in heirloom varieties these days. Your operation might be the only place they can buy a certain old strain of hollyhock, for instance.

Regardless of where you live, there are practical matters to settle before engaging in a part-time business selling plants. First, find out what the local zoning regulations are and what permits are needed. Many of the greenhouses of semi-commercial growers were already on their sites as hobby units, but their uses may be stipulated by law.

Building permits may be necessary for a temporary greenhouse or other structures. Parking availability and even the size of signs are considerations, too. In case you plan to use restricted pesticides, you will need an applicator's license (see your county Extension agent for more information).

A sales tax number obtainable from the state will allow you to buy wholesale. It's also a good idea to have a bookkeeper if you're not handy at accounting. Bear in mind the tax advantages: It is possible to depreciate cost of a greenhouse and also to take tax deductions for a business phone, office space and furniture, utilities, and other expenses related to the business.

Decorative Items

Consider using your plants to make arrangements, or to sell as cuttings for other people's arrangements. Take a class in flower arranging or wreath making. Almost everyone, including children, can master the techniques involved and find themselves happily involved in a profitable venture.

With a piece of ground of your own, you can grow such decorative items as Indian and strawberry corn, gourds, martynia, okra pods, and kaffir corn. Bundles of cornstalks, wild rose hips, bittersweet, and other beauties from the roadside are great for attracting customers. As a service to customers, garden centers, supermarkets, and department stores may be willing to set up an area for the sale of your products.

A Plant Nursery

You can use your land to grow and sell fruit plants, trees, shrubs, and vines. There are thousands of species you can grow. However, most of the commercial trade is concentrated in a limited number of varieties. Producing some of the more unusual cultivated varieties is one possibility for the small nursery. Many of these varieties are especially adapted to the soils and climate of a local area.

There is a difference between growing plants as a hobby and as a business venture. Growing plants that you find pleasure in growing, on land and in an area where the climate is best adapted for those plants, can result in a profitable small business that is enjoyable for the individual or family. But if you are not willing to grow plants that are most adapted to your soil or climate, then you cannot effectively compete in the marketplace. Your investment of capital and labor will produce a low return, showing an unsatisfactory net profit or even a loss. In such cases you are better off devoting your time to hobby gardening.

Producing nursery plants requires a lot of work. And they need special attention at various stages of growth. Yet the knowledge and skill to produce quality plants is well within the realm of possibility for the average individual.

See how the successful producers do it, talk with them and seek their advice, and read current journals and literature for recommendations based on new research.

A small nursery operated on a part-time basis does not require extensive acreage. Depending on the kinds of plants produced and the size to which they are grown, 2 to 12 acres could be adequate. Because of the great number of kinds of plants produced in nurseries, it should be easy to select plants well adapted to the land and climate of the area as well as plants you will enjoy growing.

Depending on the kind of plants and the region of the country, there is a choice between producing the plants in field rows or in containers. Containers are pots made of plastic, metal, or other material that will hold the soil or growing medium in which the roots develop.

Container production provides the opportunity for producing a higher number of plants in a limited area. A nursery of this type will probably produce smaller plants, which may be sold retail at the site to home gardeners, or sold in quantity to the larger wholesale producing nurseries, garden centers, and other retail outlets.

Licensed by States

Producing nurseries are licensed by the state department of agriculture in every state, with all nursery stock inspected for freedom from insects and diseases at least once a year. Many kinds of nursery stock must be inspected two or more times during the year before they can be marketed.

Nurseries with plants found infested with insects or diseases may be restricted from selling any plants or have portions of the nursery "tagged" for nonsale until that kind of plant or that portion of the nursery has been treated and found free of the hazardous insect or disease.

As a rule, the nursery inspector is considered a welcome counselor by nurseries. In most states the inspectors hold bachelor's degrees in a biological science, preferably entomology or plant pathology. Their specialty is identifying plant pests. For specific remedial action they will refer the nursery operator to the proper agricultural Extension specialist for recommendations.

If the nursery specializes in propagating plants, those plants may be sold as "liners" to other producing nurseries for growing to sizes salable to the general public. "Liners" are young plants grown either from seed or by rooting cuttings. Cuttings are portions of the stems of growing plants. Depending on the kinds of plants and the facilities used for protecting and sheltering them while the roots are developing, cuttings may be the tips of young shoots that are just beginning to harden, or any stage between that and woody stems.

Nursery production is unique in agriculture because plants grown in the nursery are sold complete, including the roots. Often larger plants are sold with a ball of earth around the roots in which they were growing. When that ball is wrapped in burlap or a similar material to hold the soil and root mass intact, it is called a balled and burlapped plant (B & B). If the soil and root mass is in a container, usually a fiber pot, it is referred to as a balled and potted plant (B & P).

Over the years the nursery industry has developed a system for standardizing plants to facilitate sales to nursery operators, landscape contractors, and others. In the case of B & B or B & P plants, the standards specify minimum depth and diameter measurements of the root-ball according to the height or caliper (average trunk diameter) of the plant. The American Association of Nurserymen sets the standards.

Soil Needs

Nursery crops require deep, rich, well-drained soils for developing healthy and vigorous root systems that will overcome the shock of digging and transplanting into the new site.

It is estimated that 100 to 150 tons of topsoil per acre are removed with each crop dug B & B or B & P. Smaller plants that are dug bare root remove only limited quantities of soil. However, they suffer more severe transplant shock and may have low survival rates if not carefully handled.

Nursery operators carry on soil-building programs to maintain both the organic and nutrient content of the soil at high levels. This is done through production of green-manure crops (corn, grasses, or legumes not subject to soybean cyst nematode). These crops are carefully fertilized and limed so as to produce the maximum vegetative plant growth. They are then plowed under to enrich the soil.

For both field production and container production, the nursery operator desires level land, and often grades the fields to level them when establishing a nursery. In both production practices, service roadways are maintained through the nursery. Wet areas are tiled so as to remove excess water, permitting air to penetrate to the roots. Often in refitting nursery fields, a subsoiler is used to penetrate and loosen the subsoil to encourage greater depth of root growth.

In field production nurseries that are not level, use strip- or contour-farming methods to prevent water erosion. The nursery rows are made to follow the contour of the hill, and wide strips of sod are left undisturbed between cultivated blocks of plants.

Consider growing shade trees in sod and controlling the weeds in a small area around each tree. Mow the grass periodically both to control undesirable plants and to avoid providing nesting areas for rodents near the plants. During winter when other food is not abundant, rodents feed on the bark of plants, girdling and killing them.

The length of time required to produce nursery plants for landscape use may vary from three to fifteen years. Some of the bulb and tuber plants and flowering biennial and perennial plants require only one year to produce a salable crop.

Root-Pruning

To assure that B & B and bare-root plants transplant well, many trees and shrubs are root-pruned two and more times. Root-pruning is accomplished by pulling a cutting blade, often U-shaped, beneath the plants, cutting the tips of the roots. This practice stimulates development of a branched root system. The number of times this is done depends on the kinds of plants, how long they are grown in the nursery, and the type of soil they are growing in.

Shallow, fibrous-rooted plants, such as azaleas, seldom need to be root-pruned. Deeper-rooting plants such as yews, many conifers, and the shade trees produce the most desirable root system for transplanting when they are root-pruned as often as every three years. Although root-pruning retards plant growth, plants that have been root-pruned are easier to dig. Such plants have a well-branched root system and survive transplanting shock better.

Training environmental plants to make them conform to specific shapes and sizes, or to their natural habit of growth, is a continuous process. Prune most nursery plants frequently either to stimulate branching, or to eliminate undesirable branches. Because each kind of plant has a unique habit of growth, you must be familiar with these characteristics. By knowing plant peculiarities and following proper pruning practices, you can enhance the plant's individual characteristics. Yet it is no simple task to grow a block of shade trees with straight trunks and well-formed heads, or to grow uniformly branched shrubs.

Gardeners soon learn that not all plants grow well under the same conditions. Weeping willows, maples, and some hollies grow satisfactorily even when planted in poorly drained soils. However, juniper, yews, and oak trees require well-drained soils. Azaleas, rhododendrons, and andromedas grow best in acid soils. Yews, maples, roses, and forsythia prefer soils that are only slightly acid. White birch, mountain ash, and spruce prefer colder regions while crepe myrtle, camellias, and magnolias grow better in warm climates.

These differences enable a nursery operator to specialize, based on personal preference to produce a given kind of plant, the local market, and local soils and climate. Many nursery operators specialize in growing only certain kinds of plants, such as azaleas, rhododendrons, shade trees, ground covers, dwarf conifers, or herbaceous perennials. For the nursery to be profitable, you must become proficient in growing your chosen plants economically. With specialized equipment, you can develop efficient methods and facilities for propagating and growing these plants.

Container Growing

One advantage of container growing is that the number of plants grown per acre may be several times greater than when plants are grown in nursery rows or in beds. The blend of soil and soil additives in the container medium also enables the nursery to specialize in a plant that might not be efficiently produced in nursery fields in the area.

Often container-grown plants grow faster because the blend of growing media is more suitable for that kind of plant. Also, container-grown plants can be sold any time of year regardless of weather.

Although container culture offers many advantages, there are some disadvantages. You need a dependable source of quality soil and soil additives to have a continuous supply of satisfactory potting media. Because container-grown plants must be irrigated frequently, an irrigation system with an ample supply of high-quality water is essential.

In the North, East, and Northeast, most container-grown plants must be protected from low killing temperatures during the fall and winter months. Furthermore, plants only up to a certain size can be economically grown in containers.

Weeds, diseases, and insects are as much a problem in growing plants in the nursery as they are in growing vegetables, fruits, and turf. Weeds are frequently more difficult to control in nurseries than in other crops. This is due to the wide variety of plants being grown, the slow rate of growth of most environmental plants, and the limited number of herbicides available. Some nursery plants are more sensitive to herbicides than others, making it impossible for the nursery operator to use one herbicide over the entire nursery.

Because most environmental plants tend to grow slowly, especially during the first two to three years, there is little natural weed control from shading. Perennial weeds frequently become a serious problem, especially in fields where plants must remain for five years or more. With only a limited number of herbicides to choose from, nursery operators who grow a wide variety of plants must be familiar with the tolerance of each type and the herbicides that can safely be used around them.

Since most nurseries grow many different kinds of plants, insects and diseases are controlled whenever potential problems are observed. Therefore, you must continually monitor your nursery and select the proper pesticides after the insect or disease has been properly identified. The material or materials selected must not cause damage to plants treated or to the environment.

Nursery operators use bark and wood chips, waste products from sawmills and paper manufacturers, and compost made from leaves collected by neighboring cities. They were the first to use slow-release or time-release fertilizers. These fertilizers cost more, but they are more efficient and less wasteful.

Information Trips

Consult publications of the American Association of Nurserymen. Seek out state agriculture departments and county Extension agents for advice. Since the nursery industry is varied, take every opportunity to learn from others. The Cooperative Extension in most states conducts workshops, short courses, and tours, often in cooperation with state or local nurserymen's associations. These provide an excellent opportunity to learn about the industry, talk with experienced nursery operators, and visit facilities.

Look into short programs in nursery operation at state universities. These generally are open to the public and designed to teach the practical side of managing a nursery, with opportunities for on-the-job training. There is no better way of learning the business than working in a nursery.

If you start without any formal training or nursery experience, start small. Use no more than 1 acre of land if the plants are to be grown in nursery rows, $1/2$ acre if the plants are to be grown in beds, and no more than a $1/4$ acre should be grown in containers.

Testing Water and Soil

Have your soil or potting media and water supply tested. It is simpler and less expensive to correct soils and potting mixes before the crops are planted than after they are established. An early test on the quality and availability of water may also save you considerable grief in the future. Many nursery problems have been traced back to irrigation water that was too acid, was too alkaline, or contained excess salts.

Limit the number of species you grow to a dozen for two to three years. Learn to identify these plants immediately upon receiving them, label them properly when planting, and study their growth habits and cultural requirements. Give each plant ample room in the nursery to grow. Beginning nursery operators have a tendency to crowd plants too close together because they are small.

Group together species of plants with similar soil requirements and with similar growth habits. Remember, some plants prefer growing only in mildly acid soils. Growing these plants together with other varieties in one common soil will generally result in some species growing poorly while others prosper.

Growing upright plants with low spreading plants will often reduce the quality of the low spreaders. The uprights will most often shade out the others, resulting in poorly formed, low-quality plants, more susceptible to insects and diseases.

Growing quality environmental plants requires a thorough understanding of the growing needs of many kinds of plants, and their different growth habits. To keep soils productive, follow good soil conservation practices and invest some of your earnings into building new topsoil.

To protect plants from insects and disease, provide ideal growing conditions for each kind of plant, and use the proper pesticides only when necessary. To survive in the nursery industry, you must become a taxonomist (able to identify plants), a pathologist (able to identify diseases), an entomologist (able to identify insects), and a good farmer.

Dried Flower Arrangements

Like other early Americans, your great-grandmother probably used dried flowers to conceal odors of cooking and poor plumbing. She prepared jars of potpourri (dried flower petals mixed with spices) and placed them about the house to give off fragrance when the lids were lifted. Her winter bouquet on the upright piano might have included bittersweet, painted milkweed pods, and pampas grass.

Since our ancestors' time we have discovered untold varieties of plant materials, not only for use as potpourri and winter bouquets, but for various designs used all year. It remains a satisfying enterprise to grow and collect plants to preserve, not only to decorate the house, but to sell.

You need only develop an observant eye to spot the many materials nature offers. Besides those found in your own garden, a stroll along a country road or through fields and woodlands will reveal many blossoms, foliages, seedpods, and grasses. With a little knowledge of preserving and designing, they can be put to attractive and profitable use.

Do not collect materials from parks, arboretums, and public places. Do not collect plants on the endangered or threatened species lists in your state unless you have grown them in your own garden. Avoid poison ivy with its clusters of grayish berries and its bright red three-lobed foliage in autumn.

Greens and flowers, when preserved, remain in good condition for years. In fact, dry garlands and wreaths of blue delphinium and lotus blossoms, in good color and form, were found in opened Egyptian tombs dating back to 1700 B.C.

Four simple methods of preserving are:

AIR-DRYING. Strip foliage from fresh cut flowers. Tie loosely in small bundles, and hang them upside down in a warm, darkened room until dry, one to two weeks. Store in cardboard boxes. Protect from mice, which attack seeds.

Sturdy flowers, such as zinnias, marigolds, celosia, cockscomb, strawflowers, and herbs and mints dry well using this method. Most Williamsburg-type dried arrangements are made of air-dried flowers.

GLYCERINIZING. Smash the bottom 2 to 3 inches of stems of fresh-cut magnolia and other broad-leafed evergreens, and place immediately in a solution, 4 to 5 inches deep, of two parts hot water to one to two parts glycerine or antifreeze. Add more solution as it is absorbed. Let it remain two to three weeks or until leaves have become supple and golden brown.

Glycerinize deciduous branches such as oak, dogwood, and beech in midsummer before sap stops flowing. Materials thus treated remain supple and useful for years. They are excellent alone in large arrangements, or combined with fresh flowers.

PRESSING. Press flowers and foliage between sheets of porous paper (newspaper or a telephone directory) weighted down. Drying may require several weeks. A quick method is to press with a warm iron until all moisture is removed. This produces a flat effect, unsuitable for arrangements but useful for pressed flower pictures, laminated lampshades, stationery, and other flat designs.

BURYING IN SILICA GEL. This is a commercial drying agent in which flowers dry quickly, retaining beautifully their form and color. The most delicate flowers are best preserved this way—roses, sweet peas, daffodils, and iris.

Flowers dried by this method may absorb moisture from the air, so it is advisable to use them only in the dry winter months, or in arrangements under airtight glass or clear plastic domes. Detailed directions for using silica gel are included with its purchase. A new and very quick method consists of drying flowers with silica gel in a microwave oven.

Other drying media are fine, salt-free sand or borax powder mixed half and half with white cornmeal. Lacking these, fine dry sawdust or plain cat litter can be used, but results may be disappointing.

Many materials need no special attention. Merely cut strong stalks such as corn, grains, cotton, okra, dock, sumac, goldenrod, mullein, teasel, milkweed, rose hips, bittersweet, cattails, and Queen Anne's lace, and stand upright in tall containers until ready to use. These make excellent big arrangements, alone or in combinations with fresh flowers.

Seedpods—such as tulip, daffodil, poppy, martynia, wisteria, and trumpet vine—can be clipped from their stalks and spread out to dry. Magnolia pods, sweetgum balls, sycamore buttons, acorns, nuts, small gourds, and cones should also be dried thoroughly to prevent mildew. The above-mentioned items are excellent for binding into a dry wreath.

When collecting materials, choose only those in prime condition. To obtain desired color gradations, teasels, Queen Anne's lace, and many pods and grasses can be bleached and made more effective by soaking for a short time in a solution of household bleach. (The formula is approximately $1/2$ cup of bleach to 1 gallon water, or stronger if necessary.)

Dirty material, if not too fragile, can be washed and dried thoroughly. Clean open cones by brushing vigorously between the scales with a bottle brush or a narrow paintbrush.

Herbs

Herbs are easy to grow, and increasingly easier to sell. Cautious growers can supplement their incomes selling herbs, or grow a variety of herbs for home use including cooking and herbal teas, or decoration.

With a six- to seven-month growing season, you can grow several perennial herbs that sell well at summer garden stands. Thyme, chives, balm, Corsican mint, oregano, peppermint, rosemary, savory, spearmint, and thyme all start easily in pots.

Consider transplanting herbs to hanging pots to sell. Perennials more than annuals tend to drape themselves over the edges of pots, making them especially attractive in hanging plants. Annuals like basil are less attractive to the buyer, but still attractive to the adventurous cook. Some buyers are more drawn to the decorative piece of art (pot plus plant) than to pot or plant alone. Some people will buy a hanging rosemary to look at, not to use.

Hanging pots need to be watered and drained. Pots with holes in the bottom will drain into saucers also held in the hanger. A good mixture of potting soil and vermiculite, premixed or mixed by the experienced gardener at home, is adequate. Hanging pots can be started with seeds, but you gain time by starting with rooted cuttings or plant subdivisions. Seed and plant dealers advertise in many horticultural journals. Like ground plants, hanging pots require water, light, and protection from frost. Many herbs in hanging baskets can be grown during the winter in a small greenhouse.

Eventually it helps to fertilize herbs, since constant clipping will eventually deplete the soil of nutrients. Bonemeal, well-rotted compost, weed-free manure, or a light sprinkling of 5-10-5 commercial fertilizer can be spread near but not at the base of the plant, perhaps to a radius equalling the plant's height.

Don't forget mint and ginseng. Mints offer their magic touch to herbal tea, while ginseng is a popular tea. Easy mints to grow for home consumption include apple mint, bergamot, lemon balm, lemon thyme, orangemint, peppermint, rosemary, sage, and spearmint. These are perennials that come back from the roots in spring.

If you plant mints, keep in mind that many are difficult to eliminate once established. Plan carefully before you plant perennials. Whether you plant seed or plants, determine where you want your plants to reside permanently.

To prevent herbs from spreading, set them out in sunken tins or pots, old tires, in the holes of cinder blocks—or be prepared to do battle with some of the more aggressive spreaders.

Some mints are handsome enough in their own right to be commingled with the ornamental garden. Anise-hyssop, bergamot, hyssop, and several sages (such as pineapple sage) have colorful flowers. Coleus is added to herb borders for its ornamental foliage.

Colorful foliage "varieties" have been developed in many mints. Leaves that are silver-, yellow-, and white-margined or variegated occur in the balm, sage, and thyme species. The purple foliage in various basils, bugles, and sages adds to ornamental borders. At the outer edge of the ornamental bed, creeping varieties of mint or thyme can add aroma as well as color to your walkway.

You may place favorite herbs strategically in the ornamental bed, or in a formal "knot" garden.

Land Needs

An 8- to 10-foot plot of land will supply enough herbs for a family of four. The hobbyist can get by with an 8-by-12-foot greenhouse. A perennial grower can get by with a $1/2$-acre lot and no greenhouse.

The cautious ginseng or mint grower can make some money on a small-scale farm. The best ginseng grower can survive on a few well-managed acres. The run-of-the-mill mint farmer will make less, unless the facts and/or fiction spread about ginseng could be equally spread about mints. What would happen, for example, if the world believed that bergamot, not ginseng, improved vitality and intellectual acuity and made old men young again?

Once you prepare the land, the small herb grower has no special equipment needs, but may want drying equipment. The small herb farmer will probably harvest by hand.

Herbs dried in the sun tend to lose much of their flavor and color. If you grow more mints or ginseng, choose drying racks that can provide ventilation and heat to prevent mildew during humid periods.

Good Conditions

Culinary mints include anise-hyssop, apple mint, balm, basil, bergamot, bugle, catmint, catnip, clary, horehound, hyssop, lavender, lemon basil, lemon thyme, marjoram, mint, mother-of-thyme, orangemint, oregano, peppermint, rosemary, sage, savory, spearmint, and thyme.

Trying to start ginseng from seeds can be very frustrating to the beginner. Even if you have purchased seeds from reliable dealers, you may have to wait eighteen months for the seeds to germinate. Seeds that have dried out before you planted them may never germinate. Most mint seeds, on the other hand, tend to germinate readily.

It's cheaper, of course, to buy seeds than plants, but time also has a value. From planting seeds to harvesting may take five to seven years with ginseng, and at least five months with most of the mints.

Those fortunate enough to have a greenhouse or a cold frame can start mint seedlings to transplant later to permanent sites.

Ginseng does poorly or dies in the ordinary greenhouse. It requires a shady situation, with highly organic forest floor soils recommended. Seeds or roots should be planted 12 to 18 inches apart in beds separated by walkways to permit weeding, mulching (with fallen leaves), and harvesting. Well-drained soils cleared of extraneous roots are recommended. Mints will grow well in most garden soils, but the higher the clay content the more likely the mints will suffer from waterlogging or disease during excessively wet seasons. As a rule, the brighter and drier the habitat, the higher the aromatic qualities of mints. Even so, shade-grown mints are quite satisfactory for fresh herbal tea.

Insect Repellers

Mints do not attract most pests. In fact, many are insect repellents. Mint species may contain repellents or insecticides, such as camphor, carvacrol, citral, citronellal, eugenol, furfural, linalool, menthol, and thymol, and fungicides such as furfural, menthol, salicylic acid, and thymol.

Diseases can wipe out mint monocultures, especially in heavy soils. Intercropped with ornamentals or vegetables, mints are less likely to suffer epidemics.

Rodents, though not fond of mints, can be pests in ginseng plantations. Catnip is said to repel rodents. Some people have planted some ginseng in empty tins, the tops and bottoms of which have been removed. These tin collars may discourage subterranean rodents from eating the roots. Buried screens such as those used for tulip bulbs might work as well or better.

Mints, like ginseng roots, should be harvested only when they can be properly dried. Mint leaves can be stripped off plants in the field, leaving the roots and stems intact, or the plants can be cut and tied into bundles.

For herbal tea during the summer, choose fresh-picked mints. As winter approaches, bag up the leaves and dry them to keep the family supplied with tea throughout the winter months. For home consumption, both ginseng and mints can be frozen.

The small grower needs to develop a rather personal market—say, a weekly farmers' market or roadside stand. Usually an enterprising herb grower can arrange to sell his wares at local hardware stores, craft shows, roadside stands, specialty shops, seed stores, or supermarkets.

Dried herbs, seeds, even living plants can be sold locally or via mail order. Some herb farmers package their wares as tea bags or even potpourris for market.

Greenhouse Gardening

The greenhouse is a specialized structure designed for growing plants year-round. A clear or translucent cover permits sunlight to enter, which heats the greenhouse during the day. When excessive sun heats the building, you need ventilation. During cold nights and most cold days, the greenhouse needs a heating system to maintain the desired temperature.

As energy costs have risen, farmers naturally look for ways to heat greenhouses with alternative energy. But this is not as simple as it might seem. Greenhouses with good insulation gather lots of heat when it's sunny, but the problems arise during storms and at night. The key to a good greenhouse is maintaining a constant temperature, so the important thing in using alternative energy such as the sun is to find a way to store that energy for later use. Solar photovoltaic panels and space heaters, for instance, can run very well when it's sunny, but to create reliable heat sources they require backup systems or excellent insulation or other technologies. For an overview and series of articles and fact sheets on comparing fuels and using alternatives, see a Web site maintained by Michigan State University at http://www.hrt.msu.edu/Energy/Default .htm. Knowledge of four experts—Erik Runkle and Matthew Blanchard of Michigan State, John Bartok of the University of Connecticut, and Ron Lacey of Texas A&M University—comes together at this site. See also the USDA's alternative energy Web sites at:

http://afsic.nal.usda.gov/nal_display/index.php?info_center=2& tax_level=2&tax_subject=281&topic_id=1366

http://www.energysavers.gov/your_workplace/farms_ranches/ index.cfm/mytopic=30006

Many of the resources there are still incomplete, showing that there's more to be done here.

After the initial investment in land and the greenhouse structure, the main expense will be for heat and labor. If you and your family are the labor force, then heat becomes the biggest expense. Other costs will include soil and growing media, fertilizer, pesticides, pots, seeds, and bulbs. The part-time greenhouse operator must develop a market for his products and skill with attention to details that result in quality plants.

Your greenhouse crops will vary, depending on where you sell them and to whom. Marketing includes selling wholesale to flower shops and garden centers or selling retail directly to the consumer.

A greenhouse should be on a site that takes advantage of full sun, provides good water drainage, and utilizes windbreaks. Electricity and a good water supply are needed. A separate building should be used to store equipment and supplies, to provide work space, and perhaps to house the heating system.

Size of the greenhouse should be well planned. If the hobby or business endeavor proves successful, the greenhouse will soon be too small. Plan the size, location, and layout to permit future expansion.

Larger greenhouses are more efficient and more economical to operate, as they cost less per square foot and the environment can be maintained more uniformly. Heating and ventilation systems are the most expensive items needed for a greenhouse. Their costs per unit area are less in larger greenhouses.

Structural Options

Many styles of greenhouse frames exist; select one that is pleasing and practical for you. The frame may be wood, steel, or aluminum. The cover can be glass, plastic sheet, or fiberglass, each available in different sizes and qualities.

A popular low-cost greenhouse is the pipe frame or curved roof style. The foundation is a series of pipes driven into the ground to support the curved roof members. Roof members may be made of steel or aluminum pipe or may be a curved truss.

The cover is a single or double layer of greenhouse-quality, ultraviolet-inhibited 6-mil (0.006-inch) thick or heavier plastic sheet. Plastic sheet is good for one or two years, depending on the material quality and weather. An air-inflated double layer of plastic film can reduce heating costs by 30 percent. For a more permanent cover, use a clear greenhouse-grade fiberglass. Fiberglass is available in several grades having service lives of a few years to perhaps twenty.

Plans are available through your county or state Cooperative Extension for wood frame greenhouses that can be covered by plastic film or fiberglass. Wood in contact with the soil should be pressure-treated or painted with a wood preservative. Copper naphthenate is a safe preservative near plants; creosote and pentachlorophenol are harmful to plants.

A good-quality greenhouse can be built with a good foundation and rigid frame, or an inexpensive greenhouse can be built with a temporary frame to give seasonal plant protection.

A glass greenhouse is the third possibility. Glass and aluminum or steel combine to make a long-lasting, beautiful greenhouse. The glass, rigid aluminum or steel frame, and a sturdy foundation make the initial investment high. However, annual maintenance is much less. While the glass greenhouse is a showplace, the beginner will find the less expensive, temporary plastic or fiberglass greenhouse well suited as a first structure.

Two additional structural options are the hotbed and cold frame. These are low-walled frames with cover to give plants protection during cool, windy spring weather. A hotbed has a heat source in the soil. A 3-by-6-foot cold frame or hotbed can be used to advantage for starting vegetable or flowering plants.

Heating Systems

The greenhouse can be heated with steam, hot water, or hot air. The system can be fired by any of the conventional fuels. The heating system should be fully automatic and as free from maintenance labor as possible.

A thermostat is used to control heater operation. The fan on a hot-air heater should be wired to run continuously in order to maintain uniform air temperature throughout the greenhouse. Two smaller heating units instead of one large one provides some insurance in case a heating unit fails. A small standby electrical generator is good to have in case of a power failure. Heating units must be vented to the outside if there are combustion gases. Provide an air inlet near the heating unit so oxygen is available for combustion.

Heater size is determined by the following equation: heater size (BTU/hr) = (total surface area in square feet) times (night temperature difference between inside and outside, degrees Fahrenheit) times (a heat loss factor). The heat loss factor is 0.7 for air-separated double plastic sheet and 1.2 for single-layer glass, fiberglass, or plastic sheet. These figures should be increased by adding 0.3 for hobby (small) greenhouses or for windy locations.

Ventilation is essential for producing good-quality plants. The temperature must not get too high, and a supply of carbon dioxide must be maintained. Ventilation can be provided by natural convection, using side and roof vents, or by mechanical means using exhaust fans and inlet louvers. Thermostats and electrical motors are used to automate ventilation.

The ventilation system must be able to change the air once each minute in a large greenhouse, and to change it one and a half times each minute in the hobby (small) greenhouse. Winter ventilation requirements are about one-quarter air change per minute. Two fans, with one having two speeds, are often used; the low speed of one fan is enough for winter. Motorized intake louvers are placed on the opposite wall.

The volume of a greenhouse is length times width times average height and is given in cubic feet. The fan rating will be in cubic feet per minute (cfm).

Shading materials such as saran cloth, movable lath-strip covering, lime and water, and diluted white latex paint are used to reduce light intensities and to cut the solar heat load in summer. Light reduction is necessary for those plants that grow best in low light.

Many plant functions are controlled by the length of day. Some plants such as petunia, China aster, or tuberous begonia naturally flower in the long days of summer (long-day plants), and others such as chrysanthemum or poinsettia flower in the short days of fall or winter (short-day plants).

Other plants such as carnation, rose, lilies, and everblooming begonia flower regardless of the day-length (day-neutral plants).

A greenhouse operator must protect short-day plants with a light-tight cover to induce flowering when days are long. Artificial light is used on long-day plants to induce blooming in winter months.

Temperature Control

A well-designed heating and ventilating system allows the greenhouse operator to maintain the most efficient and economical temperature for plant growth. Greenhouse night temperatures are generally maintained at 50 to 70 degrees Fahrenheit, depending on the kind of plant. The temperature is permitted to rise 10 to 15 degrees during the day before ventilation is started.

The effect of temperature on growth varies with plants. Seedlings of many crops are started at a warm temperature and then grown at lower temperatures. This is true of annual vegetable and flower plants germinated and started at 70 to 75 degrees Fahrenheit, grown at 65 degrees, and finished at 55 degrees.

Water

Water that is safe for drinking is appropriate to use in a greenhouse. Water from ponds and wells is fine, providing it doesn't contain excess amounts of salt. When plants are watered, apply a sufficient amount to moisten the entire volume of soil plus some that will drain through. This drainage helps prevent buildup of salts from the water or fertilizer used. Frequency of watering is determined by size of the plant, temperature, and the growing medium's ability to hold water.

Water is applied manually with a sprinkling can or hose. Spray nozzles or porous plastic tubing are used for watering cut flower crops. Trickle tubes may be placed into individual pots or plants. Such watering systems may be made automatic using a time clock switch that is set to water at designated times, or by using devices that operate based on dryness of the soil.

Capillary watering of pot plants is possible by placing them on a bed of sand kept continually moist. Recently a carpetlike mat of natural or synthetic fibers has been used in place of the sand.

Soils and Growing Media

Plants may be grown in many types of soils, soil mixtures, or mixtures of organic matter and inert materials without soil. The growing media mixture must be uniform in texture, hold sufficient water and drain well, be porous and well aerated, and pest-free. It need not have any available nutrients, as these are supplied in fertilizing.

Growing media range from fertile topsoil with no additions, to a variety of mixtures that may include sharp sand, peat, perlite, bark and wood chips, sludge, or composted leaves. When using soil, select a sandy loam or loam, preferably one containing organic matter. Be careful of soils containing herbicides, as they may damage your plants.

Sterilize soils and growing media before use to reduce the problem of soil insects, diseases, and weed seed. Steaming is most effective, but certain chemicals may do the job. Growing mixtures are prepared by the greenhouse operator or bought already prepared. Commercial mixtures are often more economical because they are sterilized, ready to use, and may even contain some fertilizer.

Proper application of fertilizers is another part of growing under greenhouse conditions. Have soil tests made of mixtures with a high proportion of soil. Mixtures without soil generally contain few available fertilizer nutrients.

Modern greenhouse procedures call for using soluble fertilizers. These are applied at the time of watering. Fertilizer in this form is available to the plant at once. Soluble fertilizers are generally of the so-called complete types, supplying nitrogen, phosphorus, and potassium. Some may contain other fertilizer nutrients. Mixtures without any soil often need the application of "minor fertilizer nutrients" in very minute quantities to supply plant needs.

Equipment for liquid fertilizer application ranges from simple devices that meter concentrated fertilizer solution into the water hose, to elaborate proportioning devices that are adjustable to specific concentrations.

Chrysanthemums

Commercially, chrysanthemums are produced year-round by regulating the day-length to encourage vegetative growth or flowering as needed. Many cultivars (cultivated varieties) are available for greenhouse culture.

Flower types vary from singles to full doubles, incurved, thread or spider types, or anemone forms. Flower size varies from less than an inch in diameter to exhibition types 8 to 10 inches or more. Plant habit varies from dwarf compact forms to tall ones suitable for cut flowers. You can sell both cut flowers and potted mums year-round.

The natural dates of the flowering of chrysanthemums range from late August and early September to late December. Cultivars are grouped into response groups based on the number of weeks of short days needed to produce flowering. Response groups range from 7 to 14 weeks long. The best mums for cut flowers and pot plants are in the nine-, ten-, and eleven-week response groups. A crop takes about sixteen to twenty weeks from planting to harvest, varying due to the season and response group.

Chrysanthemums grow best in porous, well-drained soil with a moderate level of fertility. Commercial growers use a 60-degree Fahrenheit night temperature.

Plant single-stem cut flowers 4 to 6 inches apart. Plant multistem cut flowers 7 to 8 inches apart. Top the plant when it is 4 to 6 inches tall, and allow two or three stems to develop.

Potted plants usually have three or four cuttings per 6-inch pot and are topped once shortly after they become established. Commercial producers of cuttings supply their customers with schedules of culture suitable for their area.

Poinsettias

Most people grow poinsettias as Christmas plants. Propagate them from stem cuttings from June through mid-September. Rooting under a mist system, in sand or perlite, requires twenty-one to twenty-four days. Some growers root directly in pots in a soil mixture.

In summer, you can grow poinsettias under glass that is slightly shaded to reduce temperatures. They should have full sun starting in mid-September. The poinsettias will flower by mid-December if they receive only normal daylight and a temperature of 60 to 62 degrees Fahrenheit. If growth is satisfactory and the bracts have developed good color by early December, the temperature may be lowered to 55 to 58 degrees. However, light of any kind at night, even a nearby streetlight, can delay flowering.

Growing media for poinsettias should be porous, well drained, and slightly acid with a pH of 6.0 to 6.5. Soluble liquid or slow-release fertilizers are used with the different cultivars.

Poinsettia cultivars are standard or self-branching types. Colors are the familiar red and also white, pink, or variegated. Growth-regulating chemicals may be used on early-propagated plants to prevent tall growth.

Tomatoes

Greenhouse tomatoes can be tricky, but it's possible to grow them even in the winter, for local specialty markets. Grow tomatoes in ground beds and train them upright 6 feet or more. Soils must have a slightly acid pH of 6.5, be sandy to silt loam, have good organic matter content, and be well drained. Install tile in the beds for drainage and for steam pasteurization of the soil before planting.

Avoid excessive fertilizing in November through January when light is the poorest and temperatures more difficult to control.

Greenhouse tomatoes are most economically grown as a fall crop to fruit from October to January, or as a spring crop fruiting from March to June. Fall yields of 6 to 10 pounds per plant can be obtained. A spring crop will yield 10 to 20 pounds.

Seed is sown and seedlings are ready for transplanting in three to four weeks. They are transplanted to pots and, when 3 to 4 inches tall, planted in the ground. Planting distance is 18 to 30 inches apart, giving each plant 4 to 5 square feet. Once planted and established, they are mulched to reduce soil compaction, using ground corncobs, peanut hulls, straw, or hay.

Side branches are removed as they develop to produce a single-stemmed plant. Flowers are tapped daily to ensure they are pollinated so that each flower will produce a fruit.

Tomatoes require 60- to 62-degree Fahrenheit night temperature. Make sure there is ventilation around the plants to keep the foliage dry. Humidity control is important because mature plants produce a lot of moisture.

Give special attention to controlling insect pests such as aphids, spider mites, white fly, and several leaf-eating insects. Foliage diseases, verticillium wilt, and especially virus diseases may be serious problems.

Annuals

Production of annual flower and vegetable plants for spring is an important greenhouse use in the months of January through May. Propagate these plants by sowing seeds in flats, and then transplant. Greenhouse growers use a variety of soil or growing media for this purpose. The growing medium should be slightly acid, well aerated, and able to hold water. A medium with this texture allows the seedlings to be removed at transplanting time with little root damage.

Germination time varies with each plant, but most plants do well at temperatures of 70 to 75 degrees Fahrenheit. Examples of germination times are six to eight days for marigold and zinnia, ten to fourteen days for petunia and snapdragon, fourteen to twenty-one days for begonia, browallia, and salvia, and twenty to twenty-five days for impatiens and lobelia.

Some seedlings grow more slowly than others, so consider this when deciding when to seed. Remember also the greenhouse temperature.

Transplant seedlings as soon as they can be handled easily, which may be ten to fourteen days after germination. Water them after transplanting and place in an area that is 60 to 65 degrees Fahrenheit. If growth proceeds too rapidly, lower the greenhouse temperature or move plants to a cold frame. Fertilize soon after transplanting.

Some pests to watch for are aphids, thrip, white fly, spider mite, botrytis, mildew, and, in the soil, damping off, rhizoctonia, and fusarium.

Other Crops

Forcing of spring-flowering bulbs works well in a greenhouse. Plant the bulbs in pots in October or November, using a well-drained soil. Tulip, narcissus, hyacinth, crocus, and others may be handled this way.

After potting, bury the pots under several inches of sand. An additional cover of straw or other material may be used later to retard freezing. If cold storage facilities capable of maintaining 35 to 38 degrees Fahrenheit are available, store the potted bulbs there. Such storage makes it easier to get the pots into the greenhouse for forcing as compared to pots buried outdoors.

After about a six-week rooting period, the pots may be brought into the greenhouse for forcing at 55 to 60 degrees Fahrenheit. The temperature may be raised if necessary.

Tulips handled this way will flower in four to five weeks. In order to have tulips or irises in flower in December or early January, the bulbs must have had some preplanting temperature treatments to hasten flower bud development. The time required for flowering varies between the different varieties and gets shorter as the normal flowering season approaches. Handle narcissus, hyacinth, crocus, and other bulbs as you handle tulips.

Beekeeping

No other activity can produce as much food and fun for the same invest-ment or for the few square feet required as can a beehive. Honeybees are the most abundant and common beneficial insect, and they thrive in most areas and climates. Few pets are as easy to care for as bees: They collect their own food—and enough extra to harvest.

Learn all you can about honeybees before starting a hive (also called a colony). Many excellent bee books, periodicals, and pamphlets are avail-able. County Extension offices and state land grant universities often distribute publications on beekeeping. Beekeeping classes are offered in many areas. Visit beekeepers in your area and ask for their advice and a demonstration of hive manipulation.

Prime Pollinators

Everyone who has fruit trees and gardens benefits from bees, which pol-linate the plants. Approximately a third of all the food we eat depends on honeybees, which provide about 80 percent of insect pollination in the United States.

Honeybees are a vital link in the ecology of man-disturbed areas where most native insect pollinators cannot survive in adequate num-bers. For example, bee-pollinated berries and fruits attract birds and pro-vide feed for other wildlife.

In many areas beekeepers rent their colonies to pollinate orchards and field crops. Income from this source can be significant if you move the hives to several crops. Generally, colonies simply are moved to the crop at the beginning of bloom and moved out when bloom is over. Two hives per acre is a rule of thumb. A word of caution: Be sure pesticides injurious to bees are not used during bloom, and have a simple contract to protect all parties.

Honey Sales

In most seasons a skilled beekeeper can produce up to 100 pounds of honey per hive. Honey may be sold at roadside stands and natural food stores, marketed by mail order and local advertising, or sold to commercial honey processors for domestic and foreign consumption.

Baked goods containing honey, such as cookies and cakes, stay fresh and moist for a long time. Many recipe books for cooking with honey are available. When honey is substituted for sugar it is usually necessary to reduce the amount of liquid in the recipe by $1/4$ cup for each cup of honey. Many fine beverages such as mead (honey wine) can be made. Honey is a quick-energy food and can replace candy in children's diets. It can also be used as a preservative for canned and frozen fruit.

One great advantage in hobby beekeeping is that honey can be harvested fresh from the hive and consumed immediately without further processing. Honey is at its best when eaten in the comb along with the natural beeswax. Yet it can be extracted easily from the comb in liquid form when desired.

Periodic hive inspections are necessary to ensure good health for the colony, abundant food, and a prolific queen. Aside from standard outdoor hives, bees may be kept indoors in a glass-walled observation hive to provide viewing of activities such as egg laying by the queen and communication dances by foragers. And there is no fear of stings by beginners because the hive is never opened indoors. Properly constructed observation hives are easily detachable and portable for use as living visual aids when giving talks at schools or clubs.

Getting Started

Before starting, determine if bees are permitted in your area. Most localities have reasonable restrictions on the number of hives permitted in residential areas. Unduly restrictive ordinances usually can be changed, especially when the benefits of bee pollination are documented properly.

Beginning beekeepers usually enjoy starting with new hive equipment, sold in kit form by bee supply companies that advertise in bee publications or the telephone directory. Experienced beekeepers may be willing to help you assemble and stock your first hive.

Frames of beeswax comb foundation are the "backbone" of the colony. Assemble them with great care. Paint new equipment and allow it to dry thoroughly before use.

Stock the hive in spring with commercially available packaged bees, containing 3 pounds of worker bees (approximately 12,000) and a queen. Or if you wish to stock your hive with a swarm, call your local fire and police departments and offer your swarm-catching services during the spring.

Buy established colonies from a beekeeper. Be sure the colonies are healthy, well fed, and in standard-dimension equipment with movable frames. Keeping bees in miscellaneous containers is impractical.

Feed sugar syrup to new colonies. Use equal volumes of sugar and water to supplement natural nectar supplies until all combs are constructed. Feeding also helps bees store enough food to ensure survival during the first winter. Don't expect to harvest honey the first year if the bees are required to construct new combs from foundation.

EQUIPMENT LIST

Item

Protective clothing
Bee veil
White coveralls
Work boots (optional)
Bee gloves (optional)

Tools for hive work
Hive tool
Bee smoker
Bee brush (optional)
Equipment for one hive
Hive stand (optional)
Standard bottom board
Brood chambers (standard hive bodies with frames)
Honey supers (shallow with frames)
Hive cover (need inner cover with telescope style)
Queen excluder
Wax comb foundation
Brood chambers
Sugar syrup feeder

Equipment for extracting honey
Electric uncapping knife
Honey extractor
Honey storage tank (optional)

The beginning beekeeper ideally should start with only one or two hives the first year. Backyards usually accommodate up to five hives if they are placed strategically near hedges, fences, or buildings to direct bee flight upward. Pets and children should be excluded from the immediate hive area.

Nectar and pollen from flowers of certain plant species supply the total food requirements of bees. They forage within at least 3 miles in every direction, an area of around 20,000 acres. Nectar and pollen availability is seasonal. Under most circumstances colonies produce far more honey than the 150 pounds needed annually for survival. The surplus is harvested as liquid honey or can be eaten in the natural honeycomb.

Biology of Bees

Honeybees are highly organized social insects. Each colony contains up to 50,000 worker bees (nonreproductive females), several thousand drones (males) during the spring and summer, and one queen. In nature, bees nest in cavities, such as hollow trees, crevices in rock cliffs, or the space between walls of buildings. They cluster inside the nest on a series of combs arranged side by side with just enough space between to permit free movement. Combs are made of pure beeswax that is secreted by worker bees and fashioned into precise hexagon-shaped cells.

During brood rearing, worker bees maintain a remarkably constant nest temperature (94 degrees Fahrenheit) even when the hive is covered with snow or exposed to scorching desert temperatures of up to 120 degrees. A combination of fanning and distributing tiny water droplets provide efficient evaporative cooling during hot weather. In cold weather, warming is achieved by muscular activity and clustering together.

In this controlled, dark environment, the queen lays around 1,500 eggs daily, 1 per cell. Eggs hatch after three days of incubation, producing tiny larvae that are fed by nurse worker bees for five to six days. Cells are then capped, and adult worker bees emerge from the cells twenty-one days after the eggs are laid. Immediately following, the cells are cleaned by housecleaner bees and the queen soon lays another egg, starting another brood cycle. Many cycles of brood in all ages and stages are developing from late winter to early fall.

During the first three weeks of adult life, some worker bees "guard" the hive entrance. They are alert to any disturbance that signals invasion by any animal that may be attracted to the rich food supply inside the hive.

Stinging behavior is the most misunderstood activity of bees. Guard bees normally are defensive and not aggressive as commonly believed. Stinging is most likely within a few feet of the hive entrance—after provocation.

Bees foraging on flowers away from the hive are nondefensive. Stings received away from the hive are very rare and usually provoked by such acts as grabbing, swatting, stepping or sitting on, or colliding with a flying bee. These are simple accidents, an element of all human activities.

A single honeybee can sting only once and it always dies within a few hours. Stinging away from the hive area contributes nothing to colony defense and makes no sense from a survival standpoint. Honeybee colonies, when properly managed and handled by the beekeeper, are quite safe. Honeybees are often mistakenly blamed for stings inflicted by many other kinds of insects, especially yellow jackets, wasps, and hornets.

If You Are Stung

Less than 1 percent of people are hypersensitive to bee stings and should not keep bees as a hobby. Fortunately, there is a recent medical advance for treating the problem—desensitization by an allergist using pure honeybee venom antigen.

If you are stung by a honeybee, immediately scrape the easily visible sting away with a fingernail before it has time to deliver the venom—thus greatly minimizing the effects. Then wash with water to remove the alarm odor, a chemical signal that directs guard bees to the "enemy." A typical reaction to a bee sting is minor discomfort near the sting site. An ice cube pressed over the area eases the pain.

Beekeepers enjoy their hobby without special concern for stings. Control the stinging tendency of guard bees by puffing smoke into the hive before and during examination. Wear protective clothing. Move slowly near the hive, as fast movements alert guard bees.

Experienced beekeepers examine colonies during maximum flight when many bees are out foraging and the incoming nectar places the bees in a "good mood." Never examine colonies at night, early mornings, late afternoons, during rainy weather, or other times when they are more defensive.

Managing Hives

In some areas hives can be kept in one location throughout the year. However, in many areas colonies are relocated seasonally near nectar-producing plants to maximize honey production and to "choose" special nectar and pollen sources. For example, to produce citrus honey, hives must be placed near citrus groves.

Consult local experienced beekeepers to determine the kinds and locations of plants most beneficial to bees. In most areas the small-scale beekeeper doesn't have to move hives elsewhere because enough forage

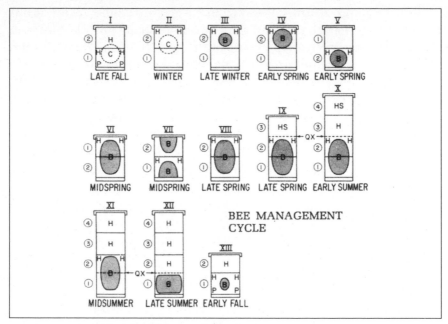

Typical honeybee colony management during seasonal changes in temperate zone areas where brood rearing ceases during late fall and midwinter. One and two are brood chambers. Three and four are honey chambers (supers) for harvested honey. *H* = stored honey; *P* = stored pollen; *C* = clustered bees in broodless nest; *B* = active brood-rearing area surrounded by clustered bees; *HS* = empty honeycombs to receive honey; *QX* = queen excluder. All views are cross-sectional, as seen from front of hive.

is usually provided by a great variety of fruit trees, ornamentals, gardens, lawns, cultivated crops, and wildflowers within foraging range.

Hives are moved best at dawn before bee flight starts. Moving hives to foraging areas is the only practical way for beekeepers to influence the selection of nectar and pollen plants. Hives must be moved at least 5 miles to prevent foraging bees from returning to the original hive location.

Efficient management is based on a thorough knowledge of bee biology and seasonal changes that affect growth of colonies. A summary of management is shown in the illustration.

By fall (I) each colony must have enough stored pollen and honey to overwinter and to support population growth during spring months, or until nectar and pollen plants bloom again. Other overwintering requirements are an adequate bee population, a young queen (less than a year old), and a protected location.

During winter in cold climates, bees cluster quietly in the broodless nest (II). Don't make inspections during this time.

In late winter, brood rearing begins (III) and the pace quickens in spring, in response to early-spring nectar and pollen, especially from fruit trees (IV). At this time the brood cluster tends to concentrate in the upper brood chamber, owing to the natural tendency for upward expansion of the brood nest.

When the upper chamber is crowded with brood (IV and VI), reverse position of the brood chambers (V and VII). This permits continued upward brood nest expansion, increases brood rearing, and diminishes excessive crowding of bees, a condition that favors unwanted swarming.

When both brood chambers become occupied with brood, add a honey chamber (super) on top, above a queen excluder, to accommodate the enlarging population and to receive any nectar and honey that may be crowding the brood chamber space (IX). The queen excluder restricts brood rearing to the lower two brood chambers.

Swarm prevention is a major problem for most hobbyists. Swarming is migration of approximately half the bees to a new nest as a means of colony reproduction. Along the way bees may cluster for a few hours or days, sometimes in very inconvenient places. Swarms cause two problems. First, they are a nuisance to people who may become frightened at the sight of thousands of flying bees, even though swarms definitely are not a sting hazard. Second, the loss of colony population greatly reduces honey production.

Swarm prevention is possible by several methods. A simple procedure is to destroy all queen cells at weekly intervals during the few critical weeks of "swarm season," usually in spring or early summer.

Throughout spring and summer, add additional honey supers as needed, usually when about half the cells in the recently added super contain nectar and/or capped honey (X). During peak production, bees may store several pounds of honey daily per colony.

Around midsummer, after the swarming season has passed and when one to two months of honey production remain, move the queen to the lower brood chamber and confine her there by a queen excluder (XI and XII). As brood emerges in the upper brood chamber (completed in three weeks), honey (50 to 60 pounds) will be stored in this chamber as food for overwintering (XIII).

This simple management program restricts brood rearing to brood combs, and honey storage to honeycombs. Honey harvested from brood combs is lower in quality. Also, if operations are timed carefully, the colony is organized for overwintering without sacrificing significant brood or honey production. Harvestable honey is easily accessible on top and can be removed by the super, rather than by sorting individual frames.

Harvesting Honey

Timing is critical in harvesting honey. You may harvest honey:

- when necessary to obtain more empty combs for storing incoming nectar;
- after a major honey flow (production peak) is completed, if you wish to separate that floral type from subsequent honeys;
- when bees are foraging actively for nectar (this avoids bee robbing or "stealing" honey from other hives because nectar is not available); and
- before the onset of cool fall weather.

You can remove bees from the honey supers by either smoking and brushing them off individual combs with a bee brush, or inserting a partition (inner cover) with a downward, one-way bee exit (for example, a Porter Bee Escape), which transfers bees from the honey super into the brood chamber in twenty-four to forty-eight hours.

These unprotected honey supers must have no openings, or robber bees will "steal" all the honey. It is important to harvest only surplus honey—leave a generous supply as bee food. Newly harvested honey supers should be placed in a warm room (85 to 100 degrees Fahrenheit) and extracted within several days. Delays are risky. Honey may granulate in the comb, it can absorb excessive moisture in humid climates, or the combs may become infested with destructive wax moths.

Basic equipment for extracting honey is an uncapping knife to cut cappings from the comb cells, a small honey extractor to remove honey centrifugally from the combs, and a small honey storage tank. Honey does not require further processing. Empty combs are stored or recycled back to the hives to be filled again.

Honey varies greatly in color, flavor, thickness, and tendency to granulate, according to the predominant floral nectar sources and conditions of storage. Most kinds of honey eventually will granulate, some within a very short time after extraction, especially if stored at the ideal temperature for granulation (50 to 60 degrees Fahrenheit).

Some granulated honeys get rock hard but when warmed to room temperature become a smooth spread of butter consistency, sometimes called "creamed" honey. Others form a coarse, "sandy" consistency that is not as pleasing to eat. The granulation process can be controlled easily.

Granulated honey can be liquefied by immersion in hot water. Minimize the time of heating to preserve quality. All honeys become darker in color and change flavor during prolonged heating and/or storage at room temperature. Long-term storage (years) is best in the freezer.

Place freshly extracted honey for home use in small containers before it granulates (keeping in mind that not all kinds of honey will granulate). Storage of honey in the refrigerator is unnecessary and usually quite inconvenient because it becomes too thick for easy pouring.

Pesticides are a major risk to bees, especially in farming areas. Avoid locating hives near frequently treated areas such as orchards, except during fruit bloom. Never store or use pesticides of any kind in the same building occupied by stored comb or wax comb foundation. Be sure to obtain current information on the protection of bees from pesticides by contacting your county agricultural agent or agricultural college.

Learn symptoms of the most serious bee diseases, especially the bacterial disease American foulbrood, which kills brood. Stored combs will be destroyed by wax moth infestation if not protected. Other dangers are floods, vandalism, theft, animals (especially bears and skunks), and fire. Keep these factors in mind when carefully choosing your hive locations.

Colony Collapse Disorder

Beginning in October 2006, some beekeepers began reporting losses of 30 to 90 percent of their hives. While colony losses are not unexpected during winter weather, the magnitude of loss suffered by some beekeepers was highly unusual.

This phenomenon, which currently does not have a recognizable underlying cause, has been termed Colony Collapse Disorder, or CCD. The main symptom of CCD is simply no or a low number of adult honey bees present but with a live queen and no dead honey bees in the hive. Often there is still honey in the hive, and immature bees (brood) are present.

The USDA's Agricultural Research Service scientists and others are in the process of carrying out research to discover the causes of CCD and develop ways for beekeepers to respond to the problem.

Researchers believe that pathogens or pesticides could be responsible for the problem. For more on this problem, see the ARS Web site http://www.ars.usda.gov/News/docs.htm?docid=15572.

Intruders and Mites

A hybrid honeybee, produced by the interbreeding of African bee stock imported into Brazil and the European honeybee, common to the U.S., became established in Brazil around 1957. In roughly thirty years, these so-called killer bees became established in tropical Latin America. According to the USDA's Agricultural Research Service, the defensive behavior and other colony characteristics of these bees strongly affected both beekeepers and the public.

Since 1990, Africanized bees have settled in the southwestern United States, the USDA found, where winters are mild. They seem especially healthy in southern Texas and Arizona. But they have not migrated east of Texas. The USDA analyzed this strange situation—why would they not travel east?—and confirmed that colder winters and higher precipitation levels along the Gulf Coast hurt these bees. It appears highly unlikely that Africanized bees will expand into Louisiana and east to Florida.

Tracheal mites, which live in the bodies of honeybees, have caused many problems for American beekeepers. The USDA found in a recent study that two-thirds of the commercial breeding colonies were resistant to the mites, which kill bees by interfering with their breathing.

USDA scientists are also studying a quarantine of Russian honeybees that migrated to North America from the former Yugoslavia, Great Britain, and eastern Russia. The aim is to see if they carry diseases or parasites that could harm the U.S. beekeeping industry.

Christmas Trees

Christmas tree production can provide supplementary income from land that otherwise might be idle. A relatively high financial return can be achieved where the emphasis is on quality trees in a quantity that can be handled by the producer.

Careful management and a commitment of time throughout the production period are necessary. During initiation of a production program, a delay of five to ten years is likely before there is a return on investment.

Though Christmas tree production can provide work during off-season periods when a landowner has time available, there are also periods during the growth of the trees when it is essential that certain cultural practices be carried out. Precision of timing can be critical in doing this work, and the trees must be observed on a regular basis. A grower should gradually develop production to a level capable of supporting a consistent marketing program.

Besides monetary gain, Christmas tree production provides other benefits. It is an excellent way to maintain, or improve, abandoned farm fields and to inhibit invasion by undesirable brush species. The relatively low growth is a valuable way of maintaining open areas as part of a desirable land-use pattern. Open areas contribute to scenic beauty and provide vantage points for scenic vistas. Such patterns contribute to recreational uses and also provide desirable habitat for many wildlife species. In some cases Christmas tree production can be combined with growing other products, but it is usually advisable to designate areas for tree production only.

Generally, an acre of suitable land can produce 700 to 900 marketable trees over an eight- to ten-year period. There is a tendency to try to produce too many trees per unit of land area. A spacing of 5 feet by 5 feet is the minimum, with wider space between the rows recommended.

Though production rates will vary according to the type of tree grown and the quality of the land, over 1,500 trees per acre within a growing period should be considered excessive. In the North, it is unlikely that more than 750 would survive.

The most efficient production and best return on investment are achieved through developing uniform plantations of trees. Always emphasize quality. Rarely have there been sufficient high-quality trees to satisfy market demands throughout the history of the Christmas tree industry.

Suitable Land

A readily accessible site is essential. Physical characteristics of the site are easiest to evaluate. More detailed analysis of soils and other environmental factors should then follow. A gentle slope, free of frost pockets, is best. Drainage of air and excessive moisture is important for good growing conditions. Direct exposure to prevailing winds, especially during winter, must be avoided.

North-facing slopes are definitely preferred over those exposed to the south. Moisture conditions are usually better on northerly slopes. Undesirable drought-prone conditions are more prevalent on south slopes where trees tend to initiate growth earlier in spring, making them more susceptible to frost damage. Though frost injury rarely kills established trees, it deforms the shapes of Christmas trees.

Evaluate soil conditions by submitting samples for testing to an appropriate laboratory in the area. Then use the relative levels of important nutrients as guides for fertilizer applications. Much can be learned from a history of the land and by observing plant cover presently growing on the site. Though you must remove competing vegetation in establishing a Christmas tree stand, its healthy condition usually indicates that trees will grow well.

Avoid areas with soil conditions that do not support good natural plant growth. Identifying the plants present on a prospective site will indicate characteristics of drainage, nutrient levels, and other aspects of soil quality. There should be sufficient soil depth to allow good development of tree roots. Consider moisture levels carefully: Excessively drained, dry sites are undesirable, and those with too much moisture should be avoided as well. Very light-textured, sandy soils with low levels of organic matter, as well as heavy clay soils, do not support good tree growth.

Also consider general weather conditions for the area. Temperature extremes, especially late-spring and early-fall frosts, are undesirable. Precipitation should occur at times that support good tree growth.

Moisture input well distributed through the growing season is important. Good winter snow cover serves to protect trees and reduce winter movement of large animals. However, winter conditions must also be considered with respect to harvesting operations because of the product's seasonal nature.

Other considerations center on accessibility and security. Christmas tree production requires ready accessibility for efficient administration of cultural practices and harvesting. Yet the final product is susceptible to theft. Have adequate security and supervision to minimize such losses.

Species Selection

Restrict your choice to species readily marketable in your region. Match the potential species against the characteristics of the land where you will plant them. Do this through a study of site requirements of the species being considered, and observe trees already growing in the area or nearby. If necessary, get advice to ensure suitability.

Once you settle on realistic candidates, evaluate them with regard to production problems. Problems include common insect and disease pests. Additional considerations are special requirements for producing quality trees of a given species, and any cultural practices that involve scheduling during a restricted season. Practices such as shearing pines only when the new growth is soft can often present problems if your personal schedule does not allow the time needed.

Species selections can usually be placed in three groups. One group includes cedars and cypresses, which have scaly and awl-shaped leaves and are produced in certain areas in the South. The remaining two groups make up most of the Christmas tree species marketed and can be classified according to needle length.

Long-needled trees are the pines and include such common Christmas tree species as Scotch pine and white pine. Pines only produce branches in annual whorls, which result in sections of bare stem along each length of annual growth. Consequently, they require a more rigid shearing schedule so as to develop desirable shape and foliage density.

Short-needled species include firs, spruces, and Douglas fir. Several are species available across our northern and central regions. The short-needled species not only have branches in annual whorls but also produce shorter branches along the length of each annual stem growth between the whorls. Because of this branching habit, shearing requirements are not as demanding as with high-quality pine trees.

Genetics is important in choosing the kinds of trees to grow. The most popular species such as Scotch pine, Douglas fir, and balsam fir exhibit a wide range of genetic variation. For efficient production of high-quality trees it is essential to have a suitable genetic strain, variety, or recognized seed source of the species to be grown. Use a genetic source specifically suited for the production site, when available.

Establishing the Stand

In some cases, improving the trees existing in natural stands on the property can provide early yields of marketable trees. Resulting income can support the establishment of plantations specifically for Christmas trees. It also allows for development of markets while the plantation trees are growing. Intensive cultural practices in natural stands can help sustain production of quality trees over long periods of time. Provide natural seed supplies to maintain regeneration of desired species. Sometimes you need to interplant to keep the area fully stocked.

Some species are suited to a practice known as stump culturing. This involves keeping a stump alive by retaining some live branches after a tree is harvested. A bud, or turned-up branch, is then cultured into a second tree on the same stump. This is usually a slower, inefficient procedure; use it only where there is no better choice.

Successful stand establishment involves achieving the highest possible percentage of trees living and developing well in their first growing seasons. This means assuring good establishment of proper stock in quantities that you can handle within the time available. Some helpful hints include:

DEVELOP AN ANNUAL PLANTING SCHEDULE aimed at sustaining an appropriate, consistent level of production.

PREPARE THE SITE BEFORE DOING ANY PLANTING. Though areas should not be exposed to erosion and other forms of deterioration, suppression or elimination of vegetation which will compete with the trees is most important.

YOU CAN MOW OR CULTIVATE, BUT SOME FORM OF HERBICIDE TREATMENT IS THE MOST EFFICIENT AND EFFECTIVE WAY. Preparation treatments will depend on the nature of the site as well as the vegetation present. Woody brush species require different treatment than vegetative cover composed of grasses and herbaceous weeds.

Safe, effective herbicides which favor Christmas tree species are usually applied as sprays. In some cases a cover crop such as rye will help hold the site in manageable condition.

EVALUATE NUTRITIONAL LEVELS IN THE SOIL PRIOR TO PLANTING as a guide to determine the need for nutrient supplements. Apply fertilizers to correct any nutrient deficiencies before planting trees.

ONCE THE SITE IS PREPARED, PLAN THE DESIRED PLANTING CAREFULLY WITH AN ADEQUATE ACCESS ROAD SYSTEM BEFORE ANY TREES ARE PUT IN. Actually mark out the planting areas and the roads beforehand. The road system is necessary for cultural operations and harvesting.

Planting Stock

Obtaining suitable planting stock can be a problem because of short supplies of good-quality material of the most desirable species. Low-priced stock usually is not the bargain it appears to be. Good stock helps assure survival and early growth, and it costs less in the long run. Evaluate stock quality carefully and inspect it before purchase if possible.

Seedlings are stock that has been grown only in the location where the seed originally germinated. Transplant stock is material that has been moved from the seedbed to an area of wider spacing to provide for a secondary period of development. Seedling stock of the pines is usually satisfactory for outplanting. Short-needled species perform much better as transplant stock.

Root systems and tops should be balanced so there is adequate root mass to support aboveground portions of the trees. A compact root system with many fine rootlets, rather than coarse, heavy roots, is desirable. The tops should have good caliper (stem diameter) and bud development, since the buds will form the basis for the first growth in the field.

Planting stock should be packaged, shipped, and handled in a way to assure maintenance of good fresh condition. Long shipping distances and exposure must be avoided. Obtain only the quantity of stock that you can handle readily in the planting operation. You can get suitable planting stock from tree nurseries. Where only seedlings of short-needled species are available, it is advisable to prepare your own transplant beds. In this way you can minimize exposure of the trees and lift them at the exact time they are needed. Two growing seasons in a properly managed transplant bed can result in excellent root and top development.

Natural Seedlings

Another possibility is to lift natural seedlings of desired species from nearby woodland and transplant them in beds adjacent to future planting sites. Don't directly plant these seedlings in your field. Nurture the natural seedlings into planting stock at relatively low cost. In the absence of recognized desirable genetic sources, this approach ensures that you have trees suitable for the local area.

Christmas tree growers can develop a home planting stock nursery. Seed collection programs—using superior trees in the area—are also possible. But management of a home nursery, although convenient and productive, is complicated and requires special knowledge and skill.

Plant the stock while it is dormant, so that new roots can grow before the first winter. There are many acceptable methods of planting. Numerous suitable hand tools are available. Spade-type tools are common and posthole augers effective. One person can plant 600 to 800 trees in a reasonable workday.

Where the terrain allows, and equipment is available, you can use tractor-drawn planting machines. In some mechanized procedures, herbicide and fertilizer applications can be combined with planting.

The planting method used should ensure good distribution of the roots since they tend to remain in their initial position during subsequent growth periods. Depth of planting should approximate the position occupied by the trees in the nursery. Pack soil firmly around the roots to eliminate all air spaces. Avoid exposure throughout the planting of young trees.

Tending the Crop

Inspect the trees frequently during the early years. Control weeds and apply fertilizer when necessary. Where healthy trees develop double tops, do early corrective trimming. Be ever watchful for early stages of development of insect and disease pests. Examine foliage, twigs, and buds for symptoms. When abnormal conditions exist, collect specimens for examination by a qualified forester. Recommended pest control treatments change frequently; obtain the most up-to-date recommendations before carrying out treatments.

As the trees begin to develop their basic Christmas tree frame, you can improve shape and density by some type of shearing treatment. Shearing practices depend on personal experience and preferences. Tools such as hedge shears and special knives are commonly used. Best results are attained when the trees are of a size that is easily reached and they are exhibiting good health and vigor.

Combinations of cultural practices produce best results. For example, best responses to shearing will occur on trees that are well nourished, free of competing vegetation, and with ample room to develop. Old practices, such as scarring the stems, produce negative responses and suppress growth, and are undesirable.

When trees of inferior quality exist in the early stages of plantation development, remove them to avoid efforts wasted on individuals that will never be marketable. Do not hesitate to cull out poor-quality, defective specimens. Produce trees in the shortest possible time to get maximum return on your investment and to minimize risk and exposure to harmful agents.

Preparing to Harvest

Locate buyers before cutting any trees. Assistance often can be obtained from other tree growers in the area or from a marketing coordinator in the local Christmas tree growers association. To sell trees most effectively, have an exact inventory of trees available for sale.

A preharvest marketing inventory should include species, size classes, an indication of relative quality, and exact location of the trees. Sale is usually made on the basis of trees cut, packaged, and collected at a truck-loading location. It is best for you to cut your own marked trees and to package them as soon as possible after cutting. Packaged trees are easier to handle, suffer less breakage, and remain in better condition.

Time of harvesting depends on weather, available labor, processing methods, number of trees to be cut, storage conditions, and marketing requirements. Keep trees cool, with some air circulation, and protected from direct wind and sunlight.

Foliage of late-cut trees and those stored completely under cover often has lower moisture content than the foliage of trees cut earlier and stored in cool outside locations. Very low temperatures at harvesting time result in brittle branches and excessive breakage.

Assemble harvested trees according to size and grade at a location readily accessible to the vehicle they will be loaded on. Growers should be familiar with the type of transportation which will be used and plan for easy loading.

When Customers Cut Their Own Trees

Another way to sell trees is to have the customers come to your farm and cut the trees themselves. This can be fun for the buyers and their families, and it will yield a higher profit for you. If you decide on this plan, you must market fewer trees, and your farm must not be too far from population centers. You can also choose to sell wreaths and other products such as maple syrup.

The sizes of trees you sell depend on market demand and can range from small table-top trees to household specimens as tall as 12 feet or more. In regions where winter conditions are not severe, there are some opportunities to sell trees with roots balled in burlap for future outplanting by consumers. Digging and root preparation require extra work, however.

As trees are harvested, strive to maintain uniformity in the tree production areas. Uniformity in size classes, species, and growth rates provides for more efficient production and eliminates injuring small trees during the harvest of larger ones.

Mark trees sold by any system before harvest. Keep records of all production practices, harvesting operations, and marketing. These records serve to guide future operations and are necessary for accounting purposes.

As you grow your trees, remember Christmas tree buyers' personal preferences. Your objective is to sustain a consistent supply of quality trees that satisfies consumer desires, while the land benefits from proper cultural practices.

Growing Nut Trees

Nut trees are a source of highly nutritious food and traditionally grown for their nut meats. They are also valuable as shade trees and as a source of food for wildlife. As a small landowner, consider growing seedlings or grafted trees to sell. Most nut tree selections, especially away from geographic areas of commercial production, are difficult to obtain. Grafting and budding trees is a skilled technique, but one you can learn.

Selling nuts is another enterprise where imagination can greatly enhance return. Selling them wholesale or at a co-op will give a relatively low return. Instead, look into direct retailing, a pick-your-own business, or mail-order sales.

Attractive packaging boosts sales and increases the dollar return. Even unfilled nuts have value when made into novelty items for sales at craft shops and church fairs.

This chapter talks about three of them: pecans, English walnuts, and chestnuts. There are many others you could grow: chestnuts, filberts, almonds (if you are in California), hickory, and more.

Pecan Orchards

The pecan tree is native to North America and commonly found in the valleys of the Mississippi River and its tributaries. Pecans also are grown in the southeastern states north of Virginia and in western Texas, New Mexico, Arizona, and California.

Georgia growers produce more pecans per year—55 million pounds—than those in any other state, even though Georgia is outside the natural range of this nut. The first commercial varieties were selections from native stands, whereas most new selections are from breeding programs.

Optimum conditions for growing pecans include deep, well-drained soils and a warm growing season. Different varieties are adapted for frost-free growing seasons of 150 to 210 days. Although the season is long enough in parts of the Northeast and Northwest to grow "northern" varieties, it is not hot enough for the kernels to develop.

Costs that prospective pecan growers need to consider include:

- Land purchase or rental
- Tractor, disk, and sprayer—if you will be using chemicals
- Irrigation, including water, leveling, and water lines
- Planting stock
- Operating costs including labor

A few other helpful components: Find a good banker, have an understanding family, and don't be afraid to work.

It's possible to confine work in a pecan orchard to weekends and holidays—spraying, fertilizing, weeding, grafting, and thinning. Pecan growing does not require vast acreages to be profitable as a vocation or hobby.

English Walnuts

English walnuts were brought to this country on English ships. They probably originated in Persia (the land that today makes up Iran and Afghanistan); hence, their alternate name, Persian walnuts. California produces 99 percent of the commercially grown English walnuts, at a rate of about 350,000 tons per year.

English walnut trees may grow very large under ideal conditions, reaching heights of 90 feet with a 60-foot spread. But in orchards they commonly achieve heights of 40 feet and spreads of 30. In home plantings, walnuts provide shade and, under proper climatic conditions, an edible crop.

Because of land value and cost of production and management, walnuts are grown intensively, under sophisticated cultural management. However, productive trees and small plantings occur throughout much of the Midwest and eastern U.S., in addition to the West Coast states.

Walnut trees require good-textured, deep, well-drained soils for optimum growth and production. Extensive root development can occur to 15 feet in deep soils.

The best sites have good air drainage and moderate temperatures. During winter, about 1,000 hours of temperature below 45 degrees Fahrenheit are needed to complete the winter-chilling requirement. At the same time, bear in mind that walnut trees can be damaged by winter temperatures as low as 14 degrees Fahrenheit, especially when followed by a warm period. However, some of the hardier Carpathian strains of English walnut regularly withstand temperatures of -20 degrees.

In the present commercial orchard areas, the growing season should have 200 or more frost-free days, but the trees will thrive and produce where the frost-free season is as short as 150 days.

Grafted varieties of proven performance are recommended instead of seedlings. Rootstocks used in California are *Juglans hindsii* and Paradox (a cross of *J. hindsii* and *J. regia*) and in the East, eastern black walnut, *J. nigra*. Nursery operators or local nut growers should be consulted as to the best variety for a given location.

Franquette is the standard old variety. It is a late-leafing and harvesting variety, with small nuts of high quality. Yields are moderate and requirements for cultural care minimal. Hartley is the preferred variety for the in-shell trade, and important for export.

Market Preferences

The domestic American market prefers packaged walnut kernels to in-shell walnuts. For this market, new varieties have been developed that have high kernel yields of good quality, such as Payne, Ashley, Serr, Vina, and Chico. These varieties bear early in the orchard's life and yield heavy crops, but in general need more cultural care than Franquette. Hanson is a productive, well-filled, small nut selection favored in the East.

Preparations before planting might include leveling the ground for irrigation, ripping the soil to eliminate restrictions in the soil profile, and fumigating to eliminate weeds, pathogens, and nematodes. Poor site preparation will result in less-than-optimum production at increased expense throughout the orchard's life. Get information on planting, irrigation, pruning, and nutrition from your state Extension service.

Walnuts are mature and ready for harvest as soon as the tissue between the kernel and the inner lining of the shell turns brown. Shortly after that time, nuts can easily be shaken from the tree. Any harvest delay will result in a darker-colored kernel, and also allow time for mold and the entrance of the navel orange worm.

You can easily harvest walnuts as they fall from the tree. It's more efficient to harvest mechanically. Special equipment required for harvest includes a tree shaker, a windrower or sweeping device to pile the nuts into a neat, long row, and a pickup machine.

Once harvested, nuts go to the huller, which mechanically removes the hulls and brushes the shells clean. From there, the nuts go into a drying facility and are treated with forced air at 109 degrees Fahrenheit for twenty-four to thirty-six hours.

Even with mechanization, harvest costs make up nearly a third of total cash costs for producing a walnut crop. The dried nuts will store for at least a year under fairly variable conditions, or up to two years if the storage temperature is kept below 40 degrees Fahrenheit.

Insects infesting fruit and wood of walnuts reduce the quality and yield of nuts. Those that directly affect the current season's crop include codling moth, walnut aphid, dusky-veined aphid, spider mite, walnut husk fly, and navel orange worm. They either cause infested nuts to drop prematurely (primarily the codling moth), or affect the nuts' quality by reducing kernel color or directly infesting the kernels, making them worthless.

Spider mites, aphids, and scale pests cause production cuts the next year. By infesting wood and leaves, they stress the tree, resulting in less fruit wood or fruit buds being produced for the next crop. These insects can be controlled by complex programs integrating either chemicals (codling moth and scale), biological control (walnut aphid), or adequate cultural practices (navel orange worm). Consult a trained pest control adviser on these matters.

Insect and disease control, perhaps more than anything else, limits the success of small plantings of walnut and other nut trees. The knowledge and equipment required for pest control, as well as other operations, often exceed the small landowner's capabilities. These limitations should be recognized before the crop is planted.

The small landowner often needs as much know-how to grow 3 acres as the grower with 3,000 acres, and expensive equipment used by the big commercial grower often is not justified with the potential return from small plantings. However, in at least some instances, ingenuity, hard work, and elimination of middlemen can give the grower with small acreage a competitive edge.

Chinese Chestnut

Of the several species of chestnut, the Chinese chestnut is the best one for nut production in the United States. It does well in areas where peaches can be grown. The American chestnut was an important timber species in the eastern U.S., but the chestnut blight fungus destroyed it in the early twentieth century.

Although numerous selections of chestnut have been named—including Crane, Eaton, Nanking, and Orrin—they are not normally available from nurseries. Thus the purchaser of trees usually has to settle for seedlings offered by mail-order nurseries. These are generally satisfactory for the homeowner, but lack the uniformity and performance that could be expected from proven varieties.

Chestnuts prefer a well-drained, acid (with a pH of 5.5 to 6.0) soil. Fertilization, cultivation, use of herbicides, and mulching are essentially the same as for other temperate climate crops. Pruning, in the early years, should only be enough to develop a single trunk and basic scaffold. Excessive pruning of young trees delays the onset of bearing.

Chinese chestnut bears at a younger age than most other nut trees, usually three to four years after transplanting. Mature trees may have to be spaced 40 feet apart—but to increase unit area yields, a 20-by-20-foot spacing initially is more practical. Just be prepared to remove trees when they begin to shade each other.

Chestnuts are fairly regular bearers compared to many of the other nut trees, which are prone to a biennial cycle. Common yields are 1 to 2 tons per acre.

The most common and ubiquitous pests are the chestnut weevils. Adult weevils lay eggs in the nuts as the kernel fills. The larvae are well developed about harvesttime. Some insecticides are capable of controlling the weevil, but none are presently registered for use. Another pest is the chestnut blight fungus; it is not generally a problem on well-grown Chinese chestnut trees.

The gall wasp was recently introduced into the U.S. and is a threat to chestnut. It destroys new shoot growth. So far the wasp has been found only in Georgia.

Chestnut trees have a place in home planting, as seen by the thousands of trees people have in their yards. Small plantings may be profitable where there is a local market for the nuts, and if problems such as squirrels and weevils can be handled.

Keeping Chickens
and Other Fowl

Backyard poultry is coming back into vogue as the local food movement shifts into higher gear. It's not unusual today for even town dwellers to have a few neighbors who have invited chickens, roosters, and other fowl to share the yard, pecking their way around, eating the kitchen scraps and providing fresh eggs, not to mention fertilizer.

A small poultry flock can provide an excellent source of fresh eggs and meat. You can sell the surplus eggs, and meat can be sold to friends and neighbors. In many locations, however, any kind of livestock or poultry is restricted or inappropriate. Check your zoning regulations.

Flock size will vary with each individual situation. To determine the correct size for you, consider the purpose of the flock, the space and time available for poultry keeping, and the market potential for surplus production. A family of four will be well supplied with eggs from a flock of ten to fifteen layers. It takes about two hens to furnish one person with one egg a day.

For meat, decide how much poultry your family will consume. You can have several different ages and kinds (broilers, turkeys, geese, etc.) so that fresh-dressed poultry is available at all times. Or you may want to raise all of them at once and put the extras in the freezer.

Some flock owners like to produce both eggs and meat for home use. The so-called dual-purpose breeds are good for this type of flock. Breeds such as Rhode Island Reds, New Hampshires, and Plymouth Rocks are good layers, and the cockerels (males) of these breeds are satisfactory for home meat production.

Getting Started

An egg-producing flock can be started in one of three ways: with hatching eggs, day-old chicks, or started pullets (young hens). Hatching eggs, besides being limited in availability, involve certain skills not required by the others. Incubation facilities and brooding equipment are needed, and there are risks and problems in obtaining and successfully hatching fertile eggs. But many people receive a certain amount of pleasure and satisfaction from hatching and rearing their own chicks.

Day-old chicks share some of the same disadvantages as hatching eggs, but are more practical. Incubation facilities are not needed, but brooding equipment is. Again, risk is involved due to the possibility of losing chicks by disease outbreaks and management problems. Also, if straight-run chicks are purchased, about half will be cockerels.

Started pullets are the most expensive (about $3.60 each) but the easiest to handle. Buy them at twenty weeks of age. A distinct advantage is that they have been vaccinated for most troublesome diseases and in general have less risks of loss from disease or mortality.

Most layers are White Leghorn or Leghorn-type crosses. Availability will be a factor in choosing. A few local hatcheries, feed stores, and mail-order farm and suburban catalogs are potential sources.

It is difficult to estimate costs for a small laying flock because people have widely varying attitudes about investing in their flock. Some will have housing for little or no cost; others will want to use their handyman skills to build an attractive unit that fits well into the landscape.

Baby Chicks

Those who decide to start with baby chicks must first locate a source and then order forty to fifty straight-run chicks or twenty-five pullet chicks. In that way, you are assured of housing fifteen to twenty pullets at twenty weeks of age. Order the chicks in advance and be fully prepared when they arrive. Spring or early-summer brooding is safest.

Electric hover-type brooders are frequently used for small flocks, although gas brooders may be used, too. They are available from several companies in various sizes. Battery-type brooders and infrared brooding lamps are satisfactory. Also, brooders may be constructed that use ordinary lightbulbs as the heat source.

Start brooders at least twenty-four hours before the chicks arrive. Adjustments can be made during this time and the house will be warm.

The temperature under the brooder at chick level should be 95 degrees Fahrenheit during the first week. Lower the temperature by 5 degrees each week until 70 degrees is reached. With infrared lamp

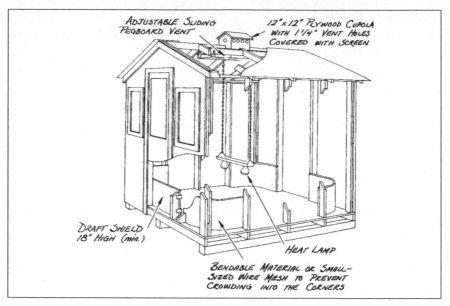

ADJUSTABLE SLIDING PEGBOARD VENT

12"×12" PLYWOOD CUPOLA WITH 1¼" VENT HOLES COVERED WITH SCREEN

DRAFT SHIELD 18" HIGH (min.)

HEAT LAMP

BENDABLE MATERIAL OR SMALL-SIZED WIRE MESH TO PREVENT CROWDING INTO THE CORNERS

An 8-by-8-foot brooder house provides enough space for raising fifty chicks or thirty-five poults (baby turkeys). Infrared brooding lamps are shown, although electric or gas hover-type brooders could be used as well. (Plans available from Poultry Science Dept., Univ. of Wisconsin, Madison, Wis. 53706)

brooding, place lamps at least 18 inches above the floor to minimize the chances of fire. If the brooder temperature is incorrect, the chicks will either huddle close to the heat source (too cold) or stay away from the heat source (too hot).

Sanitation is important in any phase of poultry production. The brooder house and all equipment should be cleaned, disinfected, and allowed to dry several days before housing chicks. Brooder houses can vary from the standard-type poultry house to the outdoor electric brooder used in warmer climates.

The brooder house has several basic requirements regardless of the type of construction. Sufficient floor space is essential; 1 square foot per chick is recommended. Ventilation should provide plenty of fresh air but prevent the chicks from being exposed to drafts. Tight construction is necessary to make the house economical to heat.

Cover the brooding area with 4 inches of absorbent-type litter (wood shavings are most commonly used). Other materials can be used, such as sawdust, peanut hulls, ground corncobs, and wood chips.

If straight-run chicks are raised, the cockerels will be ready for processing between the seventh and tenth week, depending on the size wanted and kind being reared.

Lighting

Both natural and artificial light are powerful stimulants to birds. Light influences such things as sexual maturity, behavior, and egg size. Using light effectively can help the flock owner achieve higher production and greater income. Follow the two cardinal rules: Never increase light on growing pullets, and never decrease light on laying hens.

Provide all-night lights in the brooder house for the first week. They do not need to be bright, but should provide sufficient light for the chicks to find their way around. After that, pullets being brooded between July 1 and January 1 require no additional artificial light, provided they are being raised in a house with windows.

Pullets brooded between January 1 and July 1 will be maturing when the day-length is getting longer. These flocks should be on a declining light schedule. Do this by determining the day-length when the flock will be twenty-two weeks old and adding five hours to it. Provide this amount of light for the second week. Then, reduce the total light period fifteen minutes per week, and at twenty-two weeks the pullets will be on natural day-length.

Laying birds (twenty-two weeks and older) need a constant or increasing amount of light for maximum production; normally, sixteen hours is optimum. One 40-watt incandescent bulb for 200 square feet of floor space is sufficient. To save energy, you could try compact fluorescents or LEDs for the light, but they probably would not be as hot, so another source of heat would be required. In Nigeria, agricultural researchers introduced solar chicken brooders about a decade ago. This raises the question: can't innovative homesteaders find ways other than electricity to warm their chickens and provide light? You might want to experiment with what's out there.

Laying House

The laying house does not need to be elaborate or expensive. In fact, the first time chickens are kept, the brooder house—if chicks are raised—can be converted to a laying house. Basic requirements are the same for the laying house as for the brooder house. With floor-type operations, the laying house should provide at least 2 square feet per bird for light breeds such as Leghorns, and 3 square feet for the heavier breeds.

Laying cages are excellent for a small home flock. Cages can be installed in any small building, or a house can be constructed especially for cages.

Equipment in the laying house will include feeders, waterers, nests, and lights. Besides these items, the furnishings may include containers for granite grit and oystershell, and roosts.

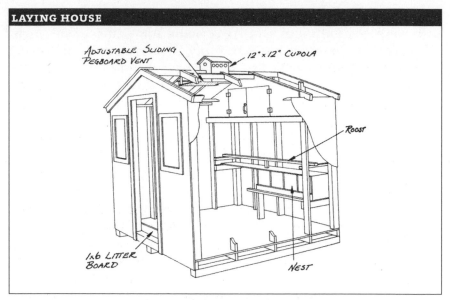

ADJUSTABLE SLIDING
PEGBOARD VENT

12"×12" CUPOLA

ROOST

1×6 LITTER
BOARD

NEST

This house is adequate for fifteen to twenty layers. Individual nests, designed to accommodate one hen at a time, are usually about a foot square. Provide a nest for each four layers. Clean, fresh water should be available at all times. With trough-type waterers, provide 1 linear inch per layer. With fountain-type waterers, provide enough fountains to have 7 to 10 gallons for each hundred layers. (Plans available from Poultry Science Dept., Univ. of Wisconsin, Madison, Wis. 53706)

The small-flock owner will probably get best results with commercial feed. No attempt should be made to get by with just table scraps and grain.

Buy prepared feeds in quantities that will be used in two or three weeks, particularly in warm weather. It takes about 19 pounds to grow a pullet to twenty weeks. Hens eat about $1/4$ pound a day, or 80 to 90 pounds a year.

Follow the manufacturer's recommendations for the particular type of feed being used. If instructions are not given, use those in the table as a guide. Feed is the major expense in producing eggs. It represents at least 60 percent of the total cost, so you should avoid feed wastage.

Don't expect every hen to lay an egg every day. Even the best flocks don't do that. But you can expect layers in a well-managed flock to lay 230 eggs per bird in a year's time. That's over 60 percent production.

Egg production is not at the same level year-round. Pullets will start laying when they are between twenty and twenty-four weeks of age. Production starts slowly and includes many small eggs. Layers "hit their peak" when they are about thirty weeks old. After that, production drops off.

Egg Handling

Eggs should be gathered at least twice daily. More frequent gathering is recommended, especially during extremely hot or cold weather. Leaving eggs in the nest increases the chances of shell damage and quality loss.

Dirty eggs should be dry cleaned or washed with an egg-washing compound as soon after gathering as possible. Use wash water that is between 100 and 120 degrees Fahrenheit. It is better to leave the eggs dirty than to wash them in cold water. Allow washed eggs to dry. Then, place them in clean cartons and refrigerate.

Producers may sell eggs from their own flocks directly to consumers. However, producers with over 3,000 hens who sell eggs to retail stores, bakeries, and restaurants are subject to federal regulations. Check with your state agriculture department about the current egg marketing laws.

The owner of a laying flock should be prepared to process his own chickens. Neither custom-dressing nor the live sale of birds is normally available.

Meat Production

A well-planned and well-managed flock can be a good source of fresh poultry meat. But large-scale commercial broiler and turkey production techniques often result in market prices difficult to match. In addition, producers are limited in the number of birds that may be sold without federal inspection. So, check with your state department of agriculture on the latest marketing regulations and make your plans accordingly.

The most economical meat production is obtained with commercial meat strains developed from breeds such as the Cornish, Plymouth Rock, and New Hampshire. These crosses have been bred for top feed conversion; they feather rapidly and mature early.

The same commercial strains of chicken meat birds can be raised for several purposes. Modern-day broilers or fryers are grown to 4 pounds in seven to nine weeks. The same bird processed at five or six weeks is marketed as a Cornish game hen; grown to twelve to fourteen weeks, it makes a delicious roaster. Males are often caponized (castrated) and sold as capons.

Broilers are brooded and reared the same as chicks raised for layers. A broiler starter ration is used. It can be fed throughout, or switched to a broiler finisher when the birds are five to six weeks old. No additional feeds or supplements are desirable or necessary. Light is not restricted on broilers. They may be raised on twenty-four hours of light for the entire growing period or placed under fourteen hours after the third week. Use twenty-four hours for the first three weeks in either case. Broiler chicks are usually available from a local hatchery, although commercial strains may not be readily available to the owner of a small flock.

Capons and Turkeys

Male chickens are caponized at three to five weeks of age to make them fatten more readily and produce higher-quality meat. If straight-run chicks are being raised, the males can be recognized from the females at this age by their larger combs, slightly larger size, and greater aggressiveness. Capons are raised to five to six months for best weight and finish. Their live weight at that time may be over 10 pounds.

The feeding and management program for capons is the same as for broilers. Use a broiler starting ration until the birds are twelve weeks old. After that, use a broiler finisher. Supply cracked corn in the afternoon after changing from the starting mash. Gradually increase the grain until the birds are getting equal amounts of corn and finishing mash at about fifteen weeks. Feed a small amount of grit once a week whenever cracked or whole corn is fed.

You can raise turkeys successfully on small farms, but they require special care and equipment. Raise turkeys away from chickens and other birds to prevent disease. The most common turkey variety is the Large White. Similar in size, but less available, is the Broad Breasted Bronze. Hens of both varieties will grow to live weights of about 15 pounds at twenty-two weeks of age; toms, 25 pounds at twenty-four weeks. Small Whites are commonly used to produce roaster-sized turkeys. At twenty-two weeks, hens will average 9 pounds and toms 14 pounds.

Start flocks with day-old poults (baby turkeys). The poults must be kept warm and dry. Feed and water the poults as soon as possible after bringing them home, dipping their beaks in water to help them learn to drink. Brood young turkeys the same as broilers.

Poults on litter need at least 1$1/_2$ square feet of floor space per bird to six weeks of age. After that, place them in a yard or larger pen, allowing at least 30 square feet per turkey. Provide a shelter outside to protect them from sun and rain.

Since turkeys are fast growing, be sure to buy correctly formulated turkey feeds. It takes about 80 pounds of feed to raise the average large turkey to market weight. For the first eight weeks, the poults need a turkey starting ration containing 26 to 28 percent protein. At eight weeks, a growing mash with 22 percent protein is recommended. Whole or cracked grain, such as corn, can be fed after eleven or twelve weeks. Equal parts of growing mash (22 percent protein) and corn (10 percent protein) provide an overall ration containing about 16 percent protein.

Ducks and Geese

People who raise small duck flocks tend to do it for the meat. The commercial duck industry is built around the Pekin breed. Pekins reach market weight in seven to nine weeks and are fairly good egg producers, but are poor setters and seldom raise a brood.

The Rouen is a popular farm flock breed. It is slower growing than the Pekin, but reaches the same weight in five to six months. Its growth rate and colored plumage make it undesirable for mass commercial production.

The Muscovy, a breed unrelated to other domestic ducks, is also used to some extent in farm flocks. It is a good forager and makes a good setter. Muscovy males are much larger than the females at market age.

Select potential breeding stock from the flock when the birds are six to seven weeks old. At that time, the sounds that ducks make can be a clue to their sex. Females have a definite sharp quack, while males have more of a muffled sound. One male to six females is recommended for breeding. Bring birds into egg production at seven months of age.

Natural methods of incubation are frequently used on small farms. Muscovy eggs require thirty-five days to hatch; all other domestic duck eggs require twenty-eight days.

Start ducklings on crumbled or pelleted chick starter, if duck feeds aren't available. Feed starter for two weeks. After that, ducklings can be fed a pelleted grower ration plus cracked corn or other grain.

Small flocks of ducklings raised in the late spring with access to green feed generally have few nutritional problems. While ducks are not as good foragers as geese, they do eat some green feed when allowed to run outdoors. Water for swimming isn't necessary for successful duck production.

Geese are excellent foragers and can live almost entirely on good pasture. Good succulent pasture or lawn clippings can be provided as early as the first week. By the time the birds are five to six weeks old, a good share of their feed can be from forage.

The Emden and Toulouse are the two most popular breeds. Emden are white in color; Toulouse vary from dark gray on their back to light gray on their breast. Normally the Emden is a much better sitter than the Toulouse. Other common breeds are the African and White Chinese.

Geese do not do well if enclosed in a house. They should be confined to a yard with a house for shelter during winter storms.

Larger breeds of geese usually mate best in pairs and trios (one male and two females). Ganders will mate with the same females year after year. Do not change these matings, unless they prove unsatisfactory. Geese two to five years of age normally give the best breeding results.

Artificial incubation can be used to hatch goslings. The incubation period varies from twenty-nine to thirty-one days. Follow manufacturer's directions for the machine being used. Goose and duck eggs require more moisture during incubation than do chicken eggs.

Provide feed and water for goslings within thirty-six hours of hatching. Feeds formulated for geese are normally not available. So, start the goslings on a crumbled or pelleted chick starter. Wheat bread or corn bread moistened with milk is a reasonable substitute for a complete ration.

After the first two or three weeks, a pelleted chick grower ration can be fed, supplemented with cracked grain. Most geese are processed for home use or sale when they are five to six months old and weigh 11 to 15 pounds.

Goose feathers are valuable, if properly cared for. As much as a pound of feathers can be plucked from three geese.

Guinea Fowl

Prime young guineas are served in homes and restaurants the same as game birds. Their flesh is tender and has a flavor resembling wild game. There are three principal varieties of guinea fowl: Pearl, White, and Lavender. The Pearl and the White are the most highly prized.

The sale of guinea hatching eggs, guinea chicks (keets), and guinea fowl for breeding is very limited. Only a few hatcheries have taken up baby keet production, although several million guineas are raised annually.

At maturity, both sexes range from 3 to $3\frac{1}{2}$ pounds in weight. Males and females differ little in appearance. Sex can be distinguished by the cry of the birds after they are about two months old. The cry of the female sounds like *buckwheat, buckwheat,* and is quite different from the one-syllable shriek of the male.

Guineas raised for breeding should have a growing diet in fall and winter prior to egg production, a breeder diet during the laying season, and a maintenance ration after laying is over. Start feeding the breeder ration, containing 22 to 24 percent protein, about a month before eggs are expected. Commercial chicken or turkey rations will give satisfactory results.

Guineas will mate in pairs, if equal numbers of males and females are present in the flock. On most general farms, however, one male is usually kept for every four or five females. The incubation period is from twenty-six to twenty-eight days. Natural methods of incubation are generally used in small flocks. For larger flocks, incubators are more satisfactory.

Chicken hens are commonly used for hatching eggs, because guinea hens will usually set on eggs only in an outdoor nest. Also, chicken hens make the best mothers for the young keets. The guinea's attachment to the chicken hen tends to control its natural wild instinct and simplifies production and management.

Keets may be raised in confinement with the same kind of equipment and brooding methods used for turkeys. They also are fed much the same as turkeys.

Guinea fowl are ready for market in fifteen to eighteen weeks. At this age, their live weight is about 3 pounds with a dressed weight of 2½ pounds. Most buyers prefer a dressed weight of at least 2 pounds.

Bantams

The word *bantam* means a small, miniature fowl. Today it may include any one of nearly 350 varieties. Most, but not all, of these miniatures are the likeness of a larger variety of domestic chicken. Generally bantams are bred for beauty of color and form rather than ability to produce eggs or meat. Bantams should be small, but don't confuse them with unhealthy, midget, or unproductive dwarf chickens. Generally adults will weigh between 16 and 30 ounces.

To get started with bantams, purchase at least a trio (one male and two females) of a breed and variety recognized by the Standard of Perfection. More than one breed or variety may be kept, but a minimum of a trio is suggested. All adult birds should be separated by varieties.

Bantams often are exhibited at shows or fairs. The awards offered, the sport of competition, and the thrill of having produced a winner all add to the excitement and benefits of the bantam show. Bantams are interesting and many are relatively unusual. They are always a popular attraction.

FEED AND WEIGHT GUIDES FOR GROWING AND LAYING LEGHORN-TYPE CHICKENS				
Age of birds	Type ration	% protein in feed	Feed consumption (lbs.) per bird for period listed	Avg. bird weight (lbs.) at end of period
0–8 wks.	Starter	20	4	1½
9–20 wks.	Grower	16	15	3¼
Laying (1 yr.)	Layer	16	80–90	4

Raising Pigs

Raising a few pigs can be interesting, fun, and a learning experience. It may also provide some income on a small scale. Pigs are very intelligent and can even become pets. However, they grow fast. Most pigs grow from about 3 pounds at birth to market weight at 225 pounds in about six months. It takes some ten months from the time the sow conceives until her pigs reach market weight.

Pigs can be sold alive at a livestock market or perhaps processed into pork for home use at a local livestock slaughtering facility. The most important products from hogs are hams, roasts, chops, bacon, and sausage.

Before acquiring pigs, get additional information from your county agricultural agent, and check on local regulations about keeping animals. The best ways of getting started raising pigs are:

- Buy a bred sow or gilt (a female that has yet to bear young) and produce a litter of pigs, then sell the litter as weaned pigs or grow them to market weight.
- Buy weaned pigs (known as feeder pigs) and feed them to market weight.

Although there are several breeds to choose from, it's best for the small operator to select crossbred animals. Crossbred sows are usually better mothers than purebreds. They farrow more pigs and faster-growing pigs. They are also more vigorous, and fewer die.

The quickest way to produce a litter is to buy a bred gilt, or an older sow that has produced one or more litters. Select sows or gilts that have twelve to fourteen well-spaced teats without deformity. Try to obtain breeding females that are themselves from litters of eight or more pigs. They should have structurally sound feet and legs. Select pigs that walk free and easy.

The pregnancy or gestation period is about 114 days. Usually eight to twelve pigs weighing about $2^{1}/_{2}$ to $3^{1}/_{2}$ pounds each are farrowed. A gilt, a young sow in her first pregnancy, usually has fewer baby pigs than older sows that have produced one or more litters.

On the average, producers lose about 15 percent of live pigs farrowed before they are weaned. Mortality from weaning to market is very low, usually less than 3 percent.

If you're raising only a few litters, it will probably not pay to buy a boar. Buy bred females or make arrangements with another swine producer to have females bred.

Another method of mating is by artificial insemination (AI). This is desirable for disease control but should be used only if good technical help is available, such as an AI technician or a producer who has had experience with AI.

When buying feeder pigs, select pigs from a reliable source where pigs are raised under sanitary conditions. Pigs should be healthy, weaned, and started on feed. Buy pigs of uniform age and size that weigh between 35 and 60 pounds. Choose females or castrated males (barrows).

Shelter and Equipment

Pigs require shelter that is dry and free of drafts. The place where they will be kept should be completely ready before you bring them home. You will need an appropriate building, a shady place in summer, a good hog-tight fence, a self-feeder or feed trough, and a waterer. A simple house can be used for swine if it keeps out drafts, snow, and rain, provides shade in hot weather, and has a dry floor.

The hog shelter may be all or part of an existing older building or a small individual house. The simplest would be an A-frame with a watertight roof that forms two sides of the building, and a rear wall. The front of such a house is usually open but can be fitted with a door. If your house is movable, face it away from the wind and don't place it where water puddles.

Keep the inside of the house dry, clean, and well bedded with straw, peanut hulls, or wood shavings. Remove the bedding when it gets wet and dirty, and spread it on a field or pasture. To avoid complaints from neighbors about unpleasant odors, do not locate hog houses or haul manure within 500 feet of your neighbors' property.

In hot weather, hogs need protection from sun and heat. Hog houses should be made so they can be opened for good ventilation. Keep hogs out of airtight structures in hot weather. Trees give good shade; however, livestock should be fenced away from valuable trees. Another method of providing shade is to place four posts in the ground, connect them at the top with a framework of poles, lumber, or wire fence, and cover with material such as straw that provides shade. The shading materials should be about 4 feet aboveground.

Hog lots must be fenced hog-tight. For larger lots of several acres, woven wire fencing (32 inches high), with a strand of barbed wire at the bottom of the fencing or just above the ground, works well. For smaller lots, temporary or permanent board fence (1-by-6-inch boards) or wire panels (about 35 inches high) will be easier to construct. Attach the boards or panels to steel or wood posts. Electrical fencing is satisfactory once pigs are trained to it.

Hogs can be fed in a trough, pan, or self-feeder. Make the trough long enough so all hogs can eat at one time. If a self feeder is used, provide a feeder hole for each four to five hogs.

Hogs should have plenty of clean water at all times. A 35-pound pig drinks about $1/2$ gallon per day; a 225-pound hog, about 1 to $1^1/2$ gallons; and a brood sow suckling a litter, about 5 gallons. You can use a heavy trough or pan that pigs cannot upset, a homemade waterer made from a steel drum, or an automatic or nipple waterer connected to a water line.

Feed Needs

Feed is the biggest expense in raising hogs, about 70 to 75 percent of the total cost of production. Swine need a balanced ration or diet each day. The complete ration, which can be purchased from a feed supplier, should contain energy, protein, vitamins, and minerals.

Corn is the standard grain (the energy source) for hogs, but barley, wheat, grain sorghum (milo), and oats also can be fed. The protein, vitamins, and minerals are provided with a complete protein supplement available from a feed supplier. The grain and protein supplement can be ground and mixed together as a complete feed, or the corn and supplement can be fed separately after the pigs weigh about 75 pounds. Another way is to buy the protein such as soybean meal, a mineral premix, and a vitamin premix separately, and then mix these with the grain source.

Whichever method you use, be sure the ration has the correct amounts of nutrients for the age of the pig being fed. Follow the mixing directions and any regulations on the feed tag. Get more information on feed sources from your county agricultural agent.

You can lower feed costs by providing a good environment and selecting animals that gain fast and efficiently. It will require $3^1/2$ to $3^3/4$ pounds of feed to produce a pound of live pork. Therefore, a hog fed a complete ration will need about 650 to 700 pounds of feed to grow from weaning weight (40 pounds) to market weight at 225 pounds.

To estimate potential profits, compare your total feed cost to the expected market price for live hogs. Besides feed costs, take into account any other production costs, such as buildings and equipment, utilities, veterinary expenses, and bedding.

Feed can be supplied in a self-feeder where hogs will have access to it at all times. Or pigs can be hand-fed all they will consume in about thirty minutes, twice a day.

Hogs do well on pasture. About $1/5$ acre of good pasture is recommended for a sow and her litter, or for three to five growing pigs. Alfalfa and ladino clover are considered the best pasture. However, red clover, alsike, white clover, and lespedeza also make good pastures for hogs. Rye, oats, wheat, cattail millet, rape, soybeans, crimson clover, and cowpeas can be used for temporary pasture.

Feed the same ration on pasture except for pregnant sows, which will require up to 30 percent less feed depending on the pasture quality. Hog rings may be used in the noses of sows and older pigs (over 40 pounds) to prevent them from rooting and destroying pasture.

Sow and Litter

During gestation, feed about 4 (summer) to 6 (winter) pounds of complete ration each day. Do not let sows and gilts get too fat. During gestation, gilts should be fed so they will gain about 75 pounds, and sows about 30 pounds.

A sow will farrow around 112 to 115 days after she is bred. On the 109th day of gestation, or about 5 days before she farrows, move her into a cleaned and disinfected farrowing house or pen.

If a farrowing crate is not used, install guardrails, if possible, to prevent the sow from lying on her pigs. Place a layer of bedding in the house or pen. Use straw, peanut hulls, or wood shavings. Remove wet bedding and manure daily to keep the pen dry.

If weather permits, wash the sow with soap and warm water before moving her into the farrowing facility. Be sure to wash her teats. Washing removes worm eggs and other organisms that infest baby pigs.

To prevent constipation in sows, add wheat bran or other bulky ingredients to the ration at a level of 15 percent, or feed the regular diet containing a tablespoon of Epsom salt or Glauber salt.

For twenty-four hours after farrowing, give the sow water but little or no feed. On the second day, start feeding about 3 pounds of feed and increase the ration each day. She should be on full feed, about 10 to 12 pounds, when the pigs are a week old. As soon as she is on full feed, the sow may be self-fed.

Normal, healthy sows and gilts usually farrow without trouble. Farrowing normally takes two to five hours. If possible, be on hand to help. Remove immediately any membranes that cover the head of newborn pigs to prevent suffocation. If a newborn pig appears lifeless, breathing can sometimes be started by rubbing or slapping its sides.

Newborn Pigs

If pigs are piling on each other or shivering and have rough hair coats, they are probably cold. When possible, use supplemental heat lamps or covers to prevent chilling. Make sure sows or pigs cannot reach the lamp or cord. Do not hang the heat source by its cord; hang it securely. Make sure it is impossible for the lamp to touch bedding or other flammable material.

After delivery, paint the navel cords, if still wet, with a tincture of iodine (U.S.P. 2 percent solution). Clip off the tips of the eight tusklike needle teeth of the pigs.

If pigs do not have access to clean soil, they will need an iron injection or iron orally during the first three days to prevent baby pig anemia. Once pigs start eating, the ration will provide enough iron.

Pigs can be weaned between four and eight weeks of age. Male pigs that are not to be sold for breeding purposes should be castrated anytime between birth and four weeks of age.

At weaning, reduce the sow ration to about 4 or 5 pounds per day. The sow will come into heat, and then she may be rebred three to six days after pigs are weaned.

Internal parasite control begins with deworming the sow before farrowing. Deworm young pigs before seven to eight weeks of age. Also control external parasites, lice, and mange. Follow all directions and heed all precautions on labels of products used.

For information on swine health and vaccinations, check with a veterinarian or your county agricultural agent.

Raising Rabbits

You can raise rabbits on a fraction of an acre. The capital investment and land required are small compared to other livestock enterprises. Although government experts do not keep an official count, each year in the United States people raise roughly six to eight million rabbits for a variety of purposes. They raise them to sell as pets, to sell to laboratories, and even to sell for their fur. During World War II, when meat supplies were low, the numbers were much higher—about twenty-four million. Laboratories use about 600,000 rabbits each year for medical experiments and product testing.

Before starting to raise rabbits, there are two points to consider—how large will your operation be and what is your market? If you intend to raise just a few rabbits to supplement the family meat supply, you can consider a unit of three or four does and a buck. One doe will produce twenty-five to fifty rabbits a year, or about 50 to 100 pounds of meat if they are raised to fryer size, more when raised to roasters. If you are interested in a part-time business, you might expect to establish a herd of 50 to 150 does.

The amount of space and cost of equipment are not great to start a small rabbitry. Breeding animals cost $15 to $25 apiece. Your investment per pen for all-wire cages, feeders, and automatic waterers is about $20 to $30. The cost of housing varies with the climate.

If you have no previous experience, start small with a buck and a few does. Before making any commitment to raise rabbits, investigate your potential markets. There are three general markets—meat, laboratory supply, and breeding stock. There also is a small market for hides and Angora rabbit wool.

Potential meat producers can obtain information on voluntary grading and inspection of rabbits by contacting the USDA, Washington, D.C., 20250. State laws governing the sale of dressed rabbits vary, and you should be aware of the laws of your state.

If you plan to sell rabbits to laboratories, you need a license from the USDA. You can only expect to sell breeding stock after you have become known as a producer of exceptional animals.

Market Income

If you want to make money selling rabbits, you will make the most by selling to breeders. Selling to laboratories requires an established reputation and the ability to supply numbers of certain types of animals at specific times.

Retail selling of dressed or live animals calls for establishing a local market. Selling to a processor is the easiest way, but will also yield the lowest return (about 50 cents a pound live weight). In selling to a processor, it is desirable to have a year-round contract.

You might want to contact the American Rabbit Breeders Association (1925 South Main Street, Bloomington, IL, 61701, http://www.arba.net). The association publishes market prices for rabbits of all sizes.

If you are selling rabbits for meat, try to sell to a local buyer to avoid expensive shipping costs. Beware of buyback schemes. These are deals in which people offer you breeding stock, usually at exceedingly high prices, and in return promise to buy back all the animals you produce. One estimate of profit from twenty does and two bucks producing five litters a years is that your net would be about $700.

Selecting a Breed

There are about thirty-eight breeds and many more varieties of rabbits raised in the United States. Breeds can be categorized by size. Mature animals of the smaller breeds weigh 3 to 4 pounds each, those of medium breeds 9 to 12 pounds, while adults of larger breeds weigh 14 to 16 pounds. Select a breed based on the purpose your rabbits will be used for.

You can obtain information on where to buy rabbits from local breeders, rabbit clubs, ads in rabbit magazines, and the directory of the American Rabbit Breeders Association.

Before attempting to sell to a laboratory, determine its needs. Check with nearby hospitals, laboratories, and health department offices to find out the type, age, and size of animals desired.

Angora rabbits are raised for their wool, which is spun into yarn used for making garments. Usually hand spinners raise their own rabbits rather than purchasing wool from rabbit breeders.

Housing Equipment

Locate the rabbitry on a site with good drainage. Check local zoning regulations first. Housing varies with the climate. In mild areas, hutches can be placed out-of-doors in shade, or provided with shade by open shed-type buildings. During very hot weather, you may need to cool the rabbits by overhead sprayers or foggers placed within the building.

In colder climates, put hutches in buildings that give protection from the prevailing winds. During stormy weather, use drop curtains or panels.

Rabbitries in the northern states have space heaters to maintain about 40 degrees Fahrenheit in areas where the young are kindled (born). Supplemental heat may also be supplied to the young by suspending lightbulbs over nest boxes. Proper ventilation is important when the animals are raised in enclosed buildings.

Some rabbit breeders in moderate climates also raise worms, because worms will consume the feces and any spilled feed, thereby eliminating odor, waste, and some labor.

Earth floors permit absorption of urine into the soil. Alternatively, concrete floors with drains will facilitate waste disposal and lend themselves to easy cleaning with a hose. Place collection pans beneath wire-bottom cages in double or triple tiers, to catch both urine and feces. Empty and clean the pans at least twice a week.

Hutches for mature rabbits are about 2 feet high and no more than $2^{1}/_{2}$ feet deep. A length of 3 feet is recommended for small breeds, 4 feet for medium, and 6 feet for large breeds. Raise rabbits on wire floors of $^{1}/_{2}$- to 1-inch mesh, to reduce potential disease problems.

All-wire hutches are best and the most expensive. You can use a combination of wood and metal. Arrange cages in single or double tiers, or even triple tiers to save space, although the triple tier makes it hard to keep cages clean. Keep a few extra cages for sick or new animals.

A nest box placed in the hutch prior to kindling will provide seclusion for the doe and protection for the litter. During cold weather, these can be insulated with bedding such as straw, wood shavings, or sugarcane waste, with two or three layers of corrugated cardboard on the sides.

Feed and Water

Rabbits may be fed from feed crocks, hoppers, or hay mangers. Feed hoppers of the proper design and size save considerable time and labor.

Watering equipment includes crocks and automatic watering systems. Use crocks with the bottom smaller than the top so that ice will not crack them. Automatic watering systems can be set up with electric heating cables to provide water during freezing weather.

The rabbit is an herbivore. It eats grains, greens, and hay. A majority of commercial rabbit raisers feed commercial rabbit pellets, which contain all the nutrients the animals need.

Dry does, herd bucks, and junior does need rations that will keep them in good breeding condition. Pregnant does and does with litters need more nutrients. Pregnant females should be allowed all they can eat unless they become overly fat, in which case limit them to 6 to 8 ounces of pellets a day.

You can substitute high-quality hay and grains such as oats, wheat, barley, rye, and ground corn. Cut hay into 4-inch lengths and supplement it with 2 ounces of grain each day. Decide whether the increased labor of obtaining and feeding these materials is balanced by the decrease in food cost.

Green feeds may cause digestive upsets. Unless you use self-feeding hoppers to save labor, give the rabbits only as much food as they can eat in a day.

Don't be surprised to see that rabbits re-ingest part of their food. They excrete two kinds of feces, one hard and one soft. They eat only the soft. The habit, known as coprophagy, starts when they are about a month old. It is normal.

Rabbits require large quantities of fresh clean water. Salt should be provided either in the pellets or from salt blocks placed in the hutch.

Breeding and Care of Young

The proper age for the first mating depends on breed and individual development. Smaller breeds develop more rapidly and are sexually mature at four to five months, medium breeds are bred at six to seven months, and giant breeds at eight to ten months. Males may mature later than females.

The average gestation period lasts thirty-one or thirty-two days. Length of the nursing period varies with systems of management. If the young nurse the mother for eight weeks, a doe can produce four litters a year.

When in heat the female will show a red coloration of the vulva and may become restless, rubbing her chin on the hutch. It is not necessary to depend on external signs to determine when a doe is to be bred. Set up a definite schedule and follow it whether the doe shows signs of being ready or not. Move the doe to the buck's hutch for service. Mating should take place almost immediately and the doe can be returned to her own hutch.

Because ovulation occurs only after copulation, some breeders leave the doe with the buck for two successive matings to ensure adequate stimulation for egg release. Other breeders prefer to remove the doe immediately after the initial mating to avoid the possibility of fighting and injury. Does may be reintroduced into the male's cage two to seven days after mating. If she rejects him, she's probably pregnant.

If you want a more scientific method of determining pregnancy, while restraining the doe's head and shoulders in your right hand, place your left hand in front of the pelvis between the hind legs and with a gentle pressure, feel to detect any marble-sized embryos.

Place a nest box in the hutch on the twentieth day after breeding. The mother will usually pull fur from her body to line it. In warm climates, an inch of bedding material such as sugarcane should be sufficient. In colder climates, additional bedding material and an electric lightbulb over the nest box will help provide warmth for the young.

Most litters are kindled at night. As soon as the doe has become quiet, inspect the litter and remove any dead or deformed young. Baby rabbits from large litters may be placed with foster mothers having smaller litters at this time. The average litter size is seven or eight.

Watch litters closely for the first few days to make sure they are being well cared for and fed. Keep records on all production does and litters so that low producers and poor mothers may be culled. Good does will continue to produce maximum-sized litters for two to three years.

Rabbit Diseases

In all cases of infection or parasite infestation, isolate affected rabbits from the rest of the herd and consult a veterinarian. Prompt and accurate diagnosis, followed by proper treatment, can save you time, labor, and money.

Rabbits are susceptible to a variety of bacterial, viral, parasitic, and fungal diseases, which may pose real problems. Young, susceptible rabbits should not be overcrowded. Where coccidiosis has been diagnosed, disinfect cages with a 10 percent ammonia solution.

Raising Beef Cattle

Editor's Note: While it might seem an unusual enterprise for twenty-first-century country dwellers to get their beef on the hoof, it is just as feasible today as it was in 1978, when, as you will read here, the USDA advised allowing 2 to 5 acres per cow. If you buy the large amount of hay and feed these animals need, you can get by with less space.

If you have pasture and hay land on your property, raising some cattle for beef can put meat on your table and a little money in your pocket perhaps more easily than any other way of farming. You can use land not suited for other purposes, and in some cases equipment and buildings that otherwise would just gather dust.

Beef cattle thrive on a wide variety of land types, forages, and pasture. A lot of small acreages are rolling and must be kept largely in grass. A beef cow herd is one means of selling this grass.

You can start small with this endeavor. Three beef cows with calves can be expected to develop into thirty females in ten years if you keep all the good heifers. Besides that, there will be some steers to sell along the way, and the family meat supply provided for with a minimum investment.

A reliable supply of water is of prime importance. If you don't have it, don't try to raise beef calves. Each cow should have at least 2 acres of good pasture. If your pastureland is covered with brush and scrub trees or rock, 5 or more acres per cow will be required. Shelter for beef cattle can be kept to a minimum.

Good hay will be needed. If hay is the only source of winter feed, a cow requires about 20 pounds of mixed hay (grass-legume) per day or about 2,400 pounds of hay during a four-month winter feeding period.

In most cases the management time spent with a cow herd is at best minimal, so it's wise to buy cows of breeds that are unlikely to have much calving difficulty. In other words, buy the traditional British beef breeds: Hereford, Angus, Shorthorn, or crosses from them. The number of crosses available from these breeds is also more plentiful.

Cattle of the so-called exotic breeds are apt to have more calving difficulties than the British breeds. Selecting cows of these breeds would not be advisable for the small part-time operator who in all likelihood will be away during calving time.

The continental breeds—such as Charolais, Limousin, and Simmental—are not as well adapted to marginal management and feeding situations. But if a small herd owner is willing to accept these facts and provide the extra management, the continental breeds can be very productive due to their additional growth and muscular ability. To obtain the best results, cross these larger and more muscular breeds with bulls of the British breeds.

Artificial breeding in a small herd makes crossbreeding more practical because it does away with the need for keeping bulls of different breeds. Crossbreeding can improve pounds of calf produced per cow by 10 to 20 percent. Whatever system you pick, crossbreeding or straight breeding, use the best bulls available.

Fencing and Housing

Plan fences carefully and build them well. Woven wire, barbed wire, a combination of these, or wooden boards make permanent fences. Farmers usually use woven wire in situations of high animal pressure or where cattle and sheep may share the same pasture. Barbed wire will control cattle under most farm conditions. The secret of good fences is having well-set corner posts to which the fence can be fastened and stretched.

A two-strand electric fence makes an ideal temporary fence. Electric fences are best used to subdivide fields for improved pasture management rather than as line fences or perimeter fences. Electric fence wires can be attached to wood or steel posts by using insulators, or fastened directly to fiberglass or plastic posts. For safety, use fence chargers with the Underwriter Laboratory (U.L.) seal. Locate gates in or near fence corners rather than in the middle of line fences, to facilitate the movement of cattle out of a field.

Beef cows need only minimum facilities and do best if kept outside under most conditions. The digestive process that takes place when forages are fed produces large amounts of heat that is used to maintain the cow's body temperature.

The critical temperature of a mature beef cow adapted to a cold environment is zero degrees Fahrenheit or below. When the temperature goes much under zero, the cow will need extra feed to furnish the energy needed to keep her warm. Weather conditions during which a beef animal will seek shelter are wind and/or cold rain. Access to a windbreak, a woodlot, or an old barn will usually take care of this need.

Have a calving pen or two under a roof where cows with new calves can be held for one or two days after the calf is delivered, until it is apparent the calf is being taken care of. And if an animal needs treatment, it can be handled or observed much more easily when confined to a pen of this type. A 10-by-10-foot pen is adequate.

One of the most serious mistakes a small-herd owner can make is to feel sympathy for the cattle and shut them up in a barn. This brings on the problem of scours and contributes to respiratory trouble.

Managing Pastures

Permanent pastures are basic to a cow-calf enterprise. Bluegrass usually makes up the forages, along with wild white clover, if it is encouraged by adding fertilizer and is kept closely grazed. A good stand of grass alone will respond to the application of nitrogen. Permanent pastures can be improved by good management, liming, applying manure and fertilizers, and seeding with pasture legumes such as alfalfa.

Good management includes mowing to control weeds, and preventing animals from overgrazing. Permanent pastures vary greatly in their carrying capacity. Some produce less than 50 pounds of beef per acre. With improvement, these pastures can be made to produce five to ten times that much.

The problem with permanent pastures is that they go dormant in July and August. During these months a few acres of improved grass-legume pasture can be invaluable. In years of short summer rainfall, feeding hay may be the only way to carry cattle over this period of short feed supply, until fall rains bring permanent pasture back so they may be grazed until early to mid-November.

Rotation pastures that are well managed, fertilized, and composed of productive grass-legume species will have a carrying capacity much higher than a permanent pasture. Alfalfa is the most commonly used legume in conjunction with grasses such as brome, orchardgrass, or fescue.

Legume-grass pastures remain productive throughout the summer. Grass pastures without legumes peak in spring and early summer and are not too productive in late summer. When alfalfa is used in mixtures, it should be rotationally grazed because it cannot withstand constant grazing pressure.

Fertilize straight-grass pasture with nitrogen, and grass-legume pasture with phosphate and potash. However, use soil tests to determine exact fertilizer requirements.

Permanent pastures such as bluegrass, brome, or fescue that have not been pastured during the late summer or fall provide much winter feed for such cattle as replacement heifers and brood cows. Winter pastures should have good natural drainage to lessen the damage from trampling in wet weather. It may be advisable to harvest the first cutting of hay from fields you plan to leave for winter pasture. If round bales are used, the second cutting can be baled and left in the field. Strip grazing these round bales will increase carrying capacity of the winter pasture.

Timber or hills provide protection from wind, and cornstalk fields located near winter pasture will furnish additional winter feed.

Grazing Cornstalks

In the Corn Belt, the animals can graze cornstalk fields during late fall and winter providing the snow does not get too deep. Two acres of good cornstalks that yielded 100 bushels of corn per acre will carry a cow for about eighty to a hundred days. Supplement this diet with a loose mineral supplement along with trace mineralized salt. Also, feed a protein supplement fortified with vitamin A at the rate of 1 pound per head per day using a 40 percent supplement as the basis of this recommendation.

Cows will graze more palatable portions of the corn plant first. They will go after whatever grain is present, followed by the leaves and husks, and then the cobs and stalks. You cannot expect to recover 100 percent of the corn plant residue by grazing. From 15 percent to 30 percent of the potential dry matter present will be recovered by the cow in grazing a stalk field. Thus, if the yield of residue dry matter is about 2 tons per acre, expect to recover 0.3 to 0.6 ton per acre of feed.

Feeding Brood Cows

During lush pasture growth, brood cows obtain nearly all the nutrients required for carrying an unborn calf or for suckling a calf. During droughts or long severe winters, however, pregnant or lactating beef cows may not get enough nutrients, thus injuring the cow, the calf, or both. At these times a small amount of good alfalfa hay is valuable to beef cows.

The cow, as a ruminant, manufactures many but not all of the specialized nutrients required for good health. Certain basic requirements must come from her feed. A brief discussion of the nutrients required and how they can be supplied follows.

The largest portion of the feed is used for energy to maintain body heat, for muscular activity, and to repair body tissue. A cow needs additional energy to nourish an unborn calf or produce milk for a suckling calf.

Furnish a mature pregnant beef cow with about 10 pounds daily of total digestible nutrients (TDN) during the winter. Since average alfalfa hay contains 50 percent TDN, a bred cow would obtain her 10 pounds of TDN from 20 pounds of hay. On the other hand, oat straw contains about 45 percent TDN. Therefore, a cow would need to consume 23 pounds to furnish the necessary energy. Weathered range pasture has about the same energy value as oat straw.

Protein is most apt to be deficient in a bred cow's ration during the winter months. A pregnant cow needs about 0.9 pound of digestible protein per day. Since average-quality alfalfa hay contains about 10 percent digestible protein, 9 pounds of average alfalfa hay would fulfill her protein requirements. Of course, additional feed would be necessary to meet her energy needs.

A cow would have to eat 150 pounds per day of oat straw or mature western prairie hay, which has only 0.6 percent digestible protein, to meet her protein requirements. But since her dry hay capacity is less than 3 percent of her live weight, a 1,000-pound cow can consume only 20 to 25 pounds of hay per day. Therefore, an oat straw, mature western prairie hay, or weathered range grass ration would be deficient in protein.

Supplementing the ration with 5 pounds of alfalfa hay or 1 pound of soybean or cottonseed meal will provide a suitable protein ration. This can be fed free choice if one part salt is mixed with two parts soybean or cottonseed meal.

Vitamins and Minerals

The cow meets all of her vitamin needs, except vitamins A and D, from what she eats and manufactures in her rumen. Since vitamin D is obtained through exposure to the sun, beef cattle rarely suffer from a vitamin D shortage. Only vitamin A is apt to be deficient.

Vitamin A deficiency lowers resistance to colds and other related infections. Cattle deficient in this vitamin often will water at the eyes sufficiently to moisten either or both sides of the muzzle. Feeding green hay or green silage is one way to provide vitamin A. Five to 10 pounds of green leafy alfalfa hay, bright clover hay, green nonlegume hay, or legume silage will provide the daily vitamin A requirements of pregnant beef cows.

If the forage fed is weather-damaged or late cut and is lacking in green color, it probably is low in vitamin A. When that's the case, vitamin A can be added to the salt-mineral mixture, or the vitamin may be injected intramuscularly at levels of 1 to 3 million International Units per head. This injection will supply the necessary vitamin A supply for about one hundred days.

In most parts of the United States, only five of the thirteen mineral elements need to be supplied for beef cattle. The rest are consumed in sufficient quantities in natural feedstuffs.

Hays and grains contain sufficient quantities of calcium and phosphorus required for bone manufacture and maintenance. However, a mineral mixture containing calcium and phosphorus should be available to the animals at all times. Either steamed bonemeal or dicalcium phosphate are excellent sources of these two elements. Ordinary feedstuffs are always deficient in salt (sodium and chlorine). Therefore, salt should be included in the free-choice mineral mixture.

Beef cows whose rations are deficient in iodine will give birth to hairless, dead calves with enlarged thyroids (called "big necks"). Iodine, the one element deficient in the Midwest in particular, can be provided in the form of iodized salt.

Feeding trace mineralized salt is cheap insurance to ensure that your cattle get all the minute mineral elements that may be deficient in your area. Cobalt, copper, iron, and iodine are known to be deficient in some areas of the country. These should be supplied. If you are in doubt about mineral deficiencies in your area, contact the county agricultural agent, who can furnish the proper information.

A free-choice mineral mixture of two parts steamed bonemeal to one part iodized salt, or one part dicalcium phosphate and one part iodized salt, is usually an adequate mineral mixture for beef cows. Along with this free-choice mineral mixture, a supply of loose trace mineralized salt should be available. Give the herd access at all times to a free-choice salt-mineral mix, in a covered feeder to protect it from the weather.

Heifers and Bulls

Growing heifers should be fed to gain from 1 to 1.5 pounds per head daily. Replacement heifers should not be fat. They require protein, minerals, and limited amounts of energy. Too much energy will make them fat.

A 500-pound growing heifer needs about 0.9 pound digestible protein and about 7.75 pounds of TDN. Ten pounds of alfalfa hay would supply her protein requirements while 15 pounds of alfalfa hay would fill her total digestible nutrient needs. Since a 500-pound heifer can eat only about 14 pounds of feed daily, feeding 4 pounds of grain and 10 pounds of alfalfa hay would be one practical way to meet the heifer's nutrient requirements.

Another typical ration for a 500-pound heifer is 2 pounds of whole oats and 2 pounds shelled corn plus 10 pounds mixed hay and free-choice minerals. If the hay contains no legume, a pound of soybean meal should be substituted for a pound of the oat-corn mixture to provide supplemental protein.

In summer, good pasture will provide all the protein and energy needed. A free-choice mineral mixture plus loose salt should be available to the growing heifer at all times.

It is common practice for the herd bull to run with the cows during the summer pasture-breeding season. Good pasture can provide all the nutrient requirements for bulls. But during late fall and early winter, the bull should be conditioned for the next breeding season. Feed young bulls more liberally than mature bulls.

The amount of concentrate fed will depend largely upon the amount of flesh the bull is carrying. For example, a mature bull in good flesh can be conditioned on good legume hay and silage. Mature bulls that are run-down and young bulls should be fed high quality roughage: 5 or 6 pounds of equal parts whole oats and shelled corn and 1.5 to 2 pounds soybean oil meal. Bulls should have plenty of exercise during the conditioning period.

Equipment and Facilities

Every beef cattle operation needs certain minimum items of equipment to function efficiently. Four basic items are so obvious it seems foolish to list them—a pencil, pocket notebook, sharp pocketknife, and an animal thermometer.

The notebook is to write down breeding dates, calving dates, health treatment, and so forth. Writing down pertinent information, even with a small herd, is better than trying to remember every detail. Use of the pocketknife is obvious, and the thermometer, next to your own powers of observation, is the best diagnostic tool you can use.

Safe, efficient, economical handling facilities should be high on the list of equipment needs. It is virtually impossible to vaccinate, pregnancy check, dehorn, or do any of the myriad management practices that need to be done without minimum handling and restraining facilities.

Plan the working corral system before starting construction. You also should have a rope halter, nose lead, obstetrical chain with two handles, syringe and needles, and some plastic sleeves. This is the most basic list of equipment and facilities needed for beef cattle management.

Herd health must be planned and constantly maintained. Each herd owner is faced with problems peculiar to his own operation. The size of the herd, physical environment, water and feed supplies, exposure to neighboring herds, and past history of diseases must be considered in disease control and prevention.

Preventing Disease

Prevention is the key to reducing disease-caused production losses. There is no better method of prevention than effective vaccines properly administered. Vaccination to prevent the following diseases is recommended: blackleg and malignant edema, leptospirosis (lepto), infectious bovine rhinotracheitis (IBR or rednose), parainfluenze 3 (PT3), bovine virus diarrhea (BVD), brucellosis (Bang's disease: contagious abortion), and vibriosis (vibrio).

The vaccinations recommended are often modified to adapt to general management, geographic, and climatic conditions. More specific modifications may be in order to adapt to conditions of particular herds. Use the services of a good local veterinarian and follow his directions to the letter. Remember, you want healthy livestock.

Raising Sheep
Part-Time

On small properties, you can raise purebred sheep, feeder lambs, or a small flock of ewes. You can sell them to sheep farmers who rely on them for wool and meat. Many part-time producers of purebred sheep are found in New England, the Midwest, and the West. Most of these flocks consist of anywhere from six to fifty sheep, and generally can be classified as hobbyist in nature.

Many producers feel that a small ewe flock more than justifies the expense and time necessary to gain additional income on a few acres. Sheep also have the ability to produce well under confined situations. This type of production works well, especially as land costs continue to rise. High production levels (200 percent lamb crop) are required to achieve a profit. You can sell lambs at a range of $100 to about $300 these days, depending on their breed.

Feeder lambs are another possibility. Many producers buy these animals to raise them to market weight. This does not require extensive labor, and can be done with limited equipment on a small acreage. You would first buy the lambs, treat them for appropriate health problems, and gradually work them to a full feed, most often dispensed in a self-feeder. This enterprise requires good marketing conditions for lambs, a source of feeder lambs, and an understanding of diseases and nutrition.

One way to feed lambs is to keep them on elevated slatted floors. The advantages include reduced space requirements, elimination of bedding, reduction of parasites, and generally increased feed efficiencies. Your initial investment, of course, is much greater.

Some small operators produce black wool for home spinning, while others produce lamb's wool for home and community use.

Many sheep producers in the Northeast feel the best lamb market is the hothouse or baby lamb market. Special breeds and out-of-season breeding is advisable to get the most lucrative price. You can sell a feeder lamb for $90 to $120.

Pros and Cons

Capital investments are relatively small for sheep. A good commercial ewe can be purchased for $30 to $125, depending on the quality. A pure-bred ewe will cost up to $2,000. The return for sheep can be high per dollar invested if you take good care of the animals.

Sheep require a minimum amount of labor. Lambing and shearing can be planned to occur when other work is not heavy. Sheep housing can and should be kept inexpensive by using available buildings. These animals consume homegrown roughages except for a small amount of grain needed at special times. Sheep also improve the fertility of the soil.

The high cost of fencing and the sometimes high feed costs, along with the nuisance of dogs and predators, are among the disadvantages of sheep production in most areas of the country. Consider using electric fencing—it costs less and can reduce dog and predator losses.

You will not succeed if you believe that sheep can produce well on scant feed and under poor management. Sheep might be willing to clean fencerows and eat weeds and brush, but they require care and need specific, high-quality feeds at critical times, such as before and after lambing. Internal and external parasites will limit profits unless controlled by good pasture management and easily administered treatments.

Marketing

Sheep production has gone down in the last thirty years in this country, due partly to changes in a federal law in 1993 that ended four decades of government support for wool producers. Then, in 2002, the government passed a new law offering low-interest loans to wool and mohair producers.

Another federal law passed in 2000 gave lamb producers four years of financial assistance to help them compete with lamb importers. While per-capita consumption of lamb has declined to about 1 pound per person, down from about 5 pounds in the 1960s, there are regions where it still sells well. For example, the greatest demand for lamb in this country appears to be in the Northeast, where higher numbers of Middle Eastern, Caribbean, and African consumers tend to live.

Marketing lamb meat and wool can be a major problem if you are planning to produce in areas where sheep numbers are sparse. Plan your production with the market in mind.

Before starting a sheep enterprise, talk with others who have sheep regarding costs and returns, markets, and management. Contact your county agricultural or regional specialist, or your Extension animal scientist at the state university. In some cases, it is helpful to have such people visit your farm and go over the possibilities with you, or you may wish to visit them.

Study your individual situation carefully before starting; then, begin in a small way and grow into the enterprise as you become better acquainted with the animals and their management.

For more information on production, contact the American Sheep Industry Association, which maintains a Web site at http://www.sheep usa.org.

Raising Dairy Cows

Editor's Note: Although dairying on a small acreage is relatively rare, it is possible. But in the last twenty-five years, small-scale dairy farming has become ever more difficult. The small farmer is up against two problems. The first is that giant production-style dairy farms, with 1,000 or more cows, now dominate the market, making it difficult to realize a profit if you want to sell some of your milk. The second is that even if all you want is milk for your family, you can never leave home unless you get someone to do the milking. This essay by University of Minnesota dairy experts Robert Appleman and Kenneth Thomas, from the original version of this book, carried some firm warnings so that no one would walk naively into this enterprise. While anyone can raise a cow—or twenty-five cows—with the right amount of time and land, consider the following advice your boot camp.

Part-time dairy farmers might milk a small herd morning and night in addition to holding a full-time job in business or industry. Or they might raise replacement females, which they sell to other dairy farmers, thereby eliminating both the labor and the equipment required in a milking operation. Other operators have a cow, or even a few cows, with the family consuming all or most of the milk produced.

A dairy enterprise may be for fun, food, or profit, but whatever the scenario, it requires intensive management, an inflexible daily labor schedule, and a higher per-unit capital investment than most other part-time farming enterprises. Since a dairy cow's milk production varies so much during the year (depending on when she last freshened or gave birth to a calf), balancing a family's relatively constant demand for milk and dairy products becomes difficult. Thus, a dairy cow as an economic source of family food may also become questionable. Therefore, this chapter focuses on the part-time dairy operation for profit.

Labor

When considering the dairy cow for profit, you need to realize that a reasonable level of production must be achieved just to cover feed and other operating expenses. Such levels of production per cow usually require management skills beyond the average abilities of younger family members. In addition, part-time farmers face the further dilemma that the dairy cow is very demanding of their scarcest resource—labor.

Dairy farms can be labor-efficient, but this usually requires a size of operation beyond the desires and capability of most part-time farmers. The comparative advantages and problems with part-time dairying are outlined in the table on page 271.

Breed Selection

The most important consideration in choosing a breed of dairy cattle to establish a dairy farm is the present and future market situation. Most areas favor breeds that produce the largest volume of milk. This undoubtedly explains why 80 to 85 percent of the dairy cattle chosen are Holsteins. There are regions, however, where special milk markets have been developed and other breeds, especially Jersey and Guernsey, are common.

Other factors to consider include:

YOUR PREFERENCE. Pay attention to your instincts. If you want to raise Jerseys because you think they are a wonderful breed, then follow your enthusiasm. It will make you a better farmer

RESALE OR SALVAGE VALUE. Larger-breed animals are usually worth more when their useful life as milk producers is ended.

SUITABILITY FOR MEAT. Half the calves are males. The surplus calves from larger breeds are generally considered superior meat producers, although tenderness and taste evaluations of meat from the smaller breeds have been outstanding.

TEMPERAMENT. When young family members must handle cows, consider this factor: Holsteins are generally superior in this trait, but all breeds are acceptable if you handle the cattle gently.

CALVING DIFFICULTY. The Jersey breed has fewer problems with difficult calvings.

Advantages	Disadvantages
Efficiency Dairy cattle are efficient in converting forage and grassland into milk.	**Confining** Cows must be milked twice daily, every day while lactating, usually about 10 of every 12 months.
Permanency Dairy enterprises will continue to be important. They flourish even in Europe where competition between crops and livestock for limited acreages of available land is much more intense than in the U.S.	**Labor** Each cow requires 60 to 80 hours of labor annually. Part-time dairy farmers often are short of time.
Stability Because it takes two years for a heifer to mature, the dairy industry is less subject to large fluctuations in market supplies and prices received. Dairy farming provides a steady, regular income.	**Feed Quality** To achieve acceptable production per cow, top-quality forage is required. This is often difficult to purchase and requires excellent management abilities to produce.
	Capital The per-unit capital investment is greater than for most other part-time farming enterprises.
Product Quality There is no satisfactory substitute for fluid milk in the diets of infants and children.	**Regulations** Milk and milk production are highly regulated and often must comply with stringent health regulations.

Dairy farming is rarely a good enterprise for the part-time farmer unless he has an abundance of family labor. Labor requirements can be somewhat reduced with mechanization, but small herds normally operated by part-time farmers seldom justify the expense of much mechanization beyond installing a milking machine and a bulk milk cooler. Even then, to keep costs down most part-time dairy farmers should be adept at purchasing good used equipment.

A typical dairy herd consists of 55 percent cows (85 percent in milk), 25 percent heifers over 10 months old, 12 percent heifers from 6 weeks to 10 months, and 8 percent calves up to 6 weeks of age.

The primary causes of unprofitable dairy operations include low production per cow, high feed costs, and low production per person-year of labor expended. Since the last factor is almost certain to apply to the part-time dairy farmer, and it may not be possible to avoid high feed costs, it is imperative that a high level of production per cow be achieved if a profit is to be realized.

A herd of Holsteins should produce at 14,000 pounds of milk per cow annually, or better, to justify a moderate-cost facility and provide the desired return on the investment of labor and capital.

Feed Needs

Forages are the basic feed for a dairy cow, supplemented with sufficient grain, protein, minerals, and vitamins to support high production. The cow eats the whole plant, usually rather high in fiber content. Examples are alfalfa hay, corn silage, and pasture.

Dairy cows usually consume forage dry matter at about 2.0 to 2.2 pounds per 100 pounds of body weight. Thus, a 1,400-pound Holstein cow will consume between 31 and 34 pounds of 90 percent dry matter hay, or its equivalent, daily.

Some part-time farmers may choose dairy cattle because they have surplus pastureland available. Lush, growing pasture that is properly fertilized and managed can be an excellent source of nutrients. But its value decreases as it matures, and trampling is a problem, resulting in nutrient waste. Rotating cattle and maintaining fences around small fields to reduce nutrient losses requires more labor and is difficult to accomplish. Unless you do it, however, you will not get a high level of production from each cow.

Concentrate mixtures (feed grains, protein supplement, minerals, and vitamins) contain less fiber and are higher in nutrient content. As production increases, the amount of concentrate mixture fed is increased to meet nutrient needs.

Forage quality varies greatly depending on when it is harvested, how it is handled, and how it is stored. If you harvest late, or weather has damaged it, you will need to buy more feed. Furthermore, you will lose some hay in the cutting and storage process.

Thus the total forage requirement (hay equivalent) for a Holstein cow and her share of the replacement youngstock may approach 8.5 tons annually. Similarly, the amount of concentrate mixture needed will approach 2.1 and 2.6 tons per cow producing at the 11,000- and 14,000-pound levels, respectively.

The kind and form of forage fed varies tremendously in different regions because of climate, topography, and soil type. Each state's land grant university, and the county Extension agent, have the expertise and publications available to provide you with specific cropping and feeding programs. Contact them before investing time and effort in planning this phase of the enterprise.

How Much Land?

Land requirements for dairy cows vary. Since forage yields may range from 1.5 to over 6 tons of hay equivalent per acre, the land required when they forage may range from 1.3 to 5.7 acres per cow, including replacement stock. An average figure frequently used in the Great Lake states region is 3 acres per cow for forages and 1 acre per cow for feed grains.

In the North, protect cows against snow, winter winds, and subzero temperatures. In the South and Southwest, provide shade to keep them cool. Provide an overhead roof in most regions to divert rainwater, keep cattle dry, and improve working conditions.

The general floor area requirements per cow are 40 to 60 square feet in the barn or shed, 20 to 30 square feet of paved feeding area, and 12 to 18 square feet of holding area.

In terms of overall barn requirements, a stall barn usually provides 80 to 90 square feet of space per cow. When cows are in loose housing, the covered area provided averages about 50 square feet with another 100 square feet of outside concrete slab for feeding and cattle movement.

In stanchion barns, stalls that are too short make it difficult to keep cows clean. Narrow stalls contribute to teat, udder, and leg injuries. Modern standards call for stall platforms 4 feet wide and about 6 feet long.

Give cows separate areas for feeding, housing, milking, calving, and maintaining their young. While dry cows and "springer" heifers—those close to calving—are sometimes kept together, separate pens are desirable for different age groups. A prime reason for this separation is the ability to monitor what they eat.

Problem Areas

Be aware of the following potential problem areas, and how to avoid them:

- **Failure to detect "heat" or time to breed.** Observe cattle to be bred for "standing heat" three or more times daily.
- **Mastitis.** Avoid these infections by installing a good milking system, use good milking procedures, and take the time required to prevent new infections by "dipping teats" with a good disinfectant and treating cows going dry (out of production).
- **Calf losses.** Feed colostrum, provide a dry, well-ventilated calf pen, and observe calves frequently (three or more times a day).
- **Poor nutrition.** Harvest the forage at the right stage of maturity and feed enough concentrate mixture to encourage top production.
- **Poor genetics.** Use genetically superior AI (artificial insemination) sires and cull the poor producers.
- **Disease.** Watch your cows regularly, feed balanced rations, keep housing facilities clean and dry, and practice good sanitation.

Production of quality milk begins on the farm. Each producer is generally regulated by the Grade A Pasteurized Milk Ordinance of the U.S.

Public Health Service. Similar but less stringent regulations apply to manufacturing-grade milk producers.

Grade A milk must meet specific requirements (different markets vary) for bacteria, inhibitors (antibiotics), somatic cell count (mastitis), adulteration (added water), and rate of cooling and temperature of holding milk on the farm.

Production facilities require an appropriately constructed and maintained milkroom or milkhouse, barn and milking area, and a potable water supply. Routine inspection is required.

Some states permit the sale of raw milk directly to consumers, provided the consumer purchases the milk at the farm where the milk was produced. Part-time farmers located in a densely populated area occasionally sell milk in this manner to enhance their income potential. Some risk is involved.

It is advisable to pasteurize your milk. Before this process was invented, many diseases were attributed to consumption of raw milk. Pasteurizing milk destroys any disease-producing bacteria that might be present. Another benefit is that shelf life (storage time) of milk is increased by destroying any spoilage bacteria in the milk.

Home pasteurization of milk can be done in several ways. One process requires heating the milk to at least 145 degrees Fahrenheit and holding it continuously at or above this temperature for at least thirty minutes. Another process requires a temperature of at least 160 degrees and holding the milk continuously at or above this temperature for at least fifteen seconds.

Be sure that containers used to store pasteurized milk have been thoroughly sanitized to prevent recontamination of the milk. Whether milk produced on the farm is consumed entirely by the family or sold directly to another consumer, it is advisable to pasteurize it.

Selling Heifers to Others

A part-time farmer with a limited supply of labor and capital, but an excess of land suitable for forage production, may wish to raise replacement heifers for neighboring dairy farmers. This approach has special appeal when the part-timer is in a dairy area and is living on a farm with surplus buildings.

To sell replacement stock, two types of contracts are in general use. In one, you agree to raise replacement heifers for another farmer for a set period and at a specified monetary rate, which is either in terms of dollars per month or cents per pound of gain.

The second type of contract includes an option-to-purchase clause, in which you buy calves to raise into heifers, and the other farmer retains the first option to buy back any animals before they freshen (have calves of their own).

The amount of forage required to raise replacement heifers from six weeks to freshening at twenty-four to twenty-eight months is between 4.5 and 5.8 tons of hay equivalent (2.5 tons per year). This means that about 3.4 heifers can be grown out on the forage needed to maintain each cow and her share of the replacements. Thus, seventeen "springer" heifers can be raised each year on the same land that will support only ten cows. At the same time, the labor required to feed and care for these animals will be reduced by one-third, to about 50 or 55 hours annually.

The main disadvantages of raising dairy heifers are that you have to spend a couple of years putting money into them before you get a return, and you must observe the heifers in heat, which takes time and effort, so that you can breed them at the right time—or use a bull.

The Big Questions

A dairy farm operator needs a wide variety of skills. Experience is the best teacher in both animal husbandry and management decision making. If you want to try part-time dairy farming, ask yourself the following questions:

- Why do I want to be a dairy farmer?
- Can I work with these animals and get adequate production from them?
- Am I willing to work long and inflexible hours?
- Can I obtain good animals?
- Can I get sufficient capital to begin the operation?

Raising Dairy Goats

Dairy goats provide a source of refreshing and nourishing food that can be prepared in many ways: milk to drink, to make into an endless variety of cheeses and cultured products such as buttermilk and yogurt, to churn into butter, and, with the addition of a few extra ingredients, to make ice cream and candy.

But a milking goat demands attention if she is to provide food for the table. Someone has to care for the animals twice a day, seven days a week, all year long. Even if the doe is dry she must be fed and watered. When she is giving birth it may mean a trip to the barn every hour through the night to see how she is progressing.

There is an occasional sick animal to care for, perhaps a frozen water pipe to thaw, manure to clean out. There are also educational meetings to attend, to help you do a better job of management. Income from a goat project is closely aligned with the producing ability of the doe, the feeding and care provided by the owner, and the local market for the milk not needed for family use. For each doe producing over 3,000 pounds of milk per year, there are hundreds that produce under 500 pounds.

If each milking doe averages 4 pounds of milk per day and it sells for about $6, you have your $6. But you must factor in the cost of 5 pounds of hay and $1^1/_2$ pounds of grain per adult per day. Consider also the feed cost. Add the cost of upkeep on buildings, fencing, fertilizer for the pasture, plus feeding equipment, milking utensils, taxes, insurance and veterinary care, plus an occasional animal lost to death.

The short answer to all this is that you will find expenses nearly equaling income. Do not view the small dairy goat herd as a profit maker. Regard it instead as a break-even source of nutritious food for the family that can be produced on otherwise idle land, and as a delightful mental diversion that all hobbies should provide.

There are by-products of the hobby, too. Goat meat is good roasted or barbecued, and the hide makes beautiful gloves and jackets.

Breeding Stock

Selecting breeding stock is not a simple task even for an experienced breeder. There are visual signs that forecast an animal's capabilities. Is she big enough for her age? Compare her to other animals the same age in the herd or in other herds. Are her eyes bright, the hair coat smooth and soft? Do the ribs arch out and downward from the backbone; does she have greater depth in the rear rib area than in the chest area? The spring and depth of ribbing is evidence of body capacity that is so necessary for good forage intake.

View the udder when it is full of milk. It should have full, strong attachments at the body wall and not hang greatly below the hock area. Size of udder is not a reliable indicator of milk production. View the udder when milked out. It should have a collapsed appearance. Feel it to be sure there are no hard lumps. It should be soft, thin-skinned, and pliable.

Check milk production records. A youngster must be evaluated through its mother's production plus any records available on other daughters of her sire. Place a greater emphasis on records of the dam and sire than on the grandparents. Look for yearly production information rather than daily production. Weigh these records against a 1,500-pound yearly milk record produced over a 305-day period.

Dairy Herd Improvement records are better guides than daily milk weights. Shy away from animals that milked 10 pounds a day when fresh but were dry in less than 200 days. If the doe you are selecting is in milk, her records speak louder than her mothers and sisters.

The goal is to obtain an animal that can produce milk over a ten-month period. A year-round milk supply is difficult to obtain due to the seasonal breeding nature of the goat. It is impossible to obtain with a few does that stand dry an exceedingly long period each year.

Breed a doe after she weighs 80 pounds or at ten to twelve months of age. Most goats show heat between August and March. Gestation takes 150 days. Twins and triplets are common. Consider artificial insemination as a viable alternative to keeping a buck. If the herd numbers less than five milking does, it will be the more economical alternative, and higher-quality sires may be more easily obtained.

Housing Space

Cold loose housing where animals are free to move about on a dry, bedded pack offers the best quarters. Warm loose housing is discouraged because of problems with ventilation. Provide 25 square feet of space per adult female housed. Buildings constructed of wood are preferred over cinder block. Masonry construction is difficult to insulate; walls are cold and often damp. Construct a three-sided building with an open side facing away from the prevailing wind.

Keep a bedded pack at least 15 inches deep in the loafing area. Heat is produced inside the pack and animals rest comfortably in these quarters even though the outside temperature drops to zero. Natural air movement will ventilate and remove moisture.

Do not attempt to keep the inside temperature warmer than outside temperatures. Doing so will cause moisture to condense on walls and ceilings, creating a humid environment that goats do not tolerate. Most respiratory problems and stress-related diseases can be traced to wet conditions and inadequate ventilation.

Provide feeding and watering devices in another area of the barn away from the bedded pack. Much hay is wasted if fed on the bedded pack, and the chance for parasite problems increase. Watering devices always offer wet spots to spoil the bedded pack.

Locate 4-by-4-foot pens along one side to house three or four kids until they reach the age of one month. Older youngstock may be housed in 6-by-8-foot pens. Buck housing must be separate and downwind from the milking animal quarters. A 6-by-8-foot shed with the open side facing south is adequate for each buck. Provide a minimum 10-by-10-foot exercise area.

Feeding

Most goat owners find machinery costs too high to warrant growing forage. They purchase both hay and grain. An acre of land will provide ample room to house up to ten milking animals plus the youngstock, as well as some pasture area.

If you grow forage, assume a need for 2,000 pounds of hay per year for each mature animal. One acre of good producing land should provide forage to feed four milking does. When calculating forage needs, allow 3 pounds of hay per 100 pounds of animal. Add an extra 5 percent for wastage. Agronomic practices, crops, and cropping systems vary greatly. For assistance, contact your county Extension agent, who knows the soils, seed varieties, and cultural practices needed for optimum production.

Goats are ruminants and require fiber in their diet. They readily consume twigs and leafy portions of trees and bushes, along with many weeds. This is acceptable food only for animals not producing milk; the strong flavors of many such foods will carry through to the milk. In addition, some of the fiber in browse is low in digestible energy and protein, and milk production will fall off on such a diet.

Feed milk animals good-quality legume or grass hays. Grain fed to the herd must provide the extra energy and protein the animal needs but does not get from forage. Therefore, with good legume hay, a 12 percent protein-grain ration will be adequate. With average-quality grass hay, an 18 percent protein level is required.

The goat will readily eat vegetable peelings, leaves of cabbage and lettuce, tops of carrots and turnips—but these are not good food for the producing doe. If top production is your goal, feed as constant a diet as possible. Sudden changes in types of food may upset digestion and reduce milk flow.

Fastidious Eaters

Goats have fastidious eating habits. They waste hay and grain if it gets soiled. Clean feed bunks and mangers daily. Locate hay feeders away from the bedded area on a concrete pad that can be scraped clean each day. The keyhole feeder is best to prevent fecal contamination and wasting of hay.

Locate watering devices away from the bedded pack and on concrete. Goats will consume greater amounts of water in winter if the water is warmed to 60 degrees Fahrenheit. Electrically heated float-type stock waterers are available.

Keep in mind several basic points for disease prevention and control:

- **Maintain herd isolation.** Raise herd replacements. If animals must be purchased, buy them as young as three days of age from locally known, disease-free sources. Insist on health charts for all animals.
- **Visitors may bring diseases to your herd.** Insist they disinfect their footwear before walking into the barn. Keep them out of feed alleys.
- **Practice sanitation at home.** Avoid carrying diseases from one animal to another by regularly cleaning and disinfecting maternity quarters and baby kid pens. Isolate sick animals.
- **Control internal parasites by sanitary feeding and housing.** Separate youngstock from older animals in barns and pastures. Practice pasture rotation and graze youngstock on new pasture whenever possible. No worm treatment program can be effective without sanitary measures that interfere with the parasite's life cycle.

- **Draft-free, well-ventilated, well-lighted buildings are needed for goats.**
- **Pastures and exercise lots should be well drained.** They also should be free from trash and sharp objects that could cause injury, especially to the teats and udder.
- **Adequate nutrition is important.** Drugs, tonics, fancy mineral mixtures, and rumen stimulatory substances are not substitutes for good food.
- **Use a veterinarian for diagnosis, treatment and advice on health problems.** Too often veterinary aid is sought only for dying animals.
- **Goats respond to gentle and patient handling.** They will not stay healthy or produce well when abused.
- **House the buck separately from the does.** Provide plenty of opportunity for exercise. Nothing is more damaging to the herd sire than to keep him tied in a dark corner of the barn.

Tools and Equipment

On a bedded pack and with very little exercise, the goat's toes grow long and the hoof shell uneven. Pruning shears and a farrier's knife are useful tools to keep the foot well manicured.

Tattooing is necessary for identification, especially if goats are shown or registered. Cost of the equipment is about $30.

Remove horns as soon as the horn button can be located. Electric dehorners are the most humane, and the cost is about $17.

Behind each horn button and toward the midline of the head is a mass of cells that produce an oily musk, a major source of odor in the buck goat. By clipping the hair, these yellow cell masses may be seen lying just under the skin. Veterinarians can remove them surgically, or the owner can burn them out at the time of dehorning.

Castrating is necessary for males kept as pets or for slaughter. A veterinarian will do it, or the owner may use an elastrator (cost, $15), Burdizzo Emasculator (cost, $55), or a sharp jackknife. Disinfect all cutting instruments.

For disinfecting the navel, use a teacup partially filled with iodine to dip both the cord and belly-button area.

Pails and Strainers

Stainless-steel utensils are unequaled in terms of durability and ease of cleaning, and if you sell your milk commercially, this may be required. Plastic may be used for home-oriented dairies.

Do not use iron, copper, brass, white metal, or worn, plated utensils and spoons in handling milk. They cause milk to oxidize and taste like cardboard.

An electric hot-water heater and a stainless-steel double sink allow utensils to be washed and sanitized as required by many milk markets, and are recommended for all producers. A drying rack for storing clean equipment between milkings is essential to good sanitation.

Pasteurizing is important to protect the quality of milk and extend its shelf life. The enzyme lipase reacts with butterfat in milk, causing it to turn rancid and develop a goaty flavor. Pasteurization temperature destroys the enzyme.

Two-gallon electric pasteurizing units may be bought. A more economical approach is to heat milk in a double boiler to 165 degrees Fahrenheit for twenty seconds. Cool immediately. If the temperature goes higher than 165 or is held for longer than twenty seconds, the milk will taste cooked.

An ice-water bath or cold running water cools more quickly than air temperature. Therefore, cool the milk before putting it in a refrigerator. Larger dairies using milk cans may purchase used water immersion-type electric coolers from dairy equipment suppliers.

Goats are creatures of habit. Milk the does within minutes of the same time each morning and evening. Wash the udder and teats before milking. Use a disposable paper towel for each animal. Dry the udder thoroughly before milking.

Clipping the hair from the belly and udder in fall and winter will aid in producing a higher-quality milk.

Cool the milk immediately to 40 degrees Fahrenheit. Store it at 34 degrees for best preservation of quality and flavor. Do not expose milk to fluorescent lights or to sunlight.

Raising Horses

Keeping a horse or two on country land is even more popular now than it was twenty-five years ago. Even if you keep no other animal, you might keep a horse. You do it for the love of horses. You probably already realize that it is unrealistic to consider profit as an incentive for owning horses on a limited, part-time basis. Liability insurance makes it almost prohibitive for even full-time stables with professional help to operate.

If you do decide to make some money off your facilities, realize that a profit is related directly to how high a volume of business you do in your riding arenas, giving classes, or otherwise supervising other riders. Most stables cater to the novice rider who has had no formal instruction and thus is a considerable risk on a horse. A person who keeps horses for public hire must become familiar with the habits, disposition, and traits of the horses. An owner who knows a particular horse is apt to be vicious may be held liable for injuries caused by that animal.

The rental horse business is fraught with risk and should be entered into only after considering suitability of the horses involved, insurance costs, availability and safety of trails, and the expertise of those who will supervise riding. Some people use personal mounts for riding instruction. In most cases a homeowner's policy will cover liability if this enterprise falls within the confines of a casual and not a full-time enterprise. However, many recently written policies do not include this type of coverage.

If you have the required skills, training horses on limited acreage as a part-time occupation can provide supplemental income—but it's not easy to make money from these part-time endeavors. Keep in mind the volume of income that would be required to cover the costs involved in order to realize a profit. Without indoor riding facilities, weather very much limits the activities in most areas of the country. Many small stable operators are in the business as a part-time occupation purely because they love to work with people and horses.

Selecting a Horse

Horse trading is an old and sometimes not-too-honorable activity. In no other industry is the point of law caveat emptor (let the buyer beware) so closely adhered to. Where no warranty is implied or given, the buyer accepts all responsibility unless illegal practices can be proved.

Most dealers in horses are honest businessmen looking for satisfied repeat customers. But too often the sale and purchase of a horse is between private parties with neither one knowledgeable enough to be aware of the horse's faults.

The novice should seek the help of an experienced equine veterinarian or horse expert when buying a horse. There are standard examination procedures to determine stable habits, temperament and disposition, and to detect disease and unsoundness. Also there are certain conformation faults that limit the usefulness of the horse and predispose it to specific unsoundness. Whether you are paying an average or considerable amount for a horse, take steps to ensure a sound investment.

The price of horses varies from slaughter prices to thousands of dollars. Check with several dealers for an idea of the price range. Horses between the ages of eight to fourteen years, well trained, gentle, healthy, and sound, make the best beginner horses. The horse should fit your needs and you should fit the horse. A height for ease of mounting and a width to fit your legs makes for a safer, more comfortable ride.

The stallion is not for any but the very experienced horseman. Some mares become difficult to handle when in season. The gelding is a castrated male and is usually more stable than the stallion or mare.

Sources of horses include:

- Public auction barn—usually the dumping ground for less desirable horses and disease problems (strictly caveat emptor).
- Horse dealers, as discussed earlier.
- Horse farm sales; some deal in trained horses, but most sell weanlings, yearlings, two-year-olds, and surplus breeding stock of one of the registered breeds.
- Casual sales, as with an outgrown pony, former 4-H mount, owner going off to college, or just lost interest; this sometimes includes dumping an unsuitable or fractious horse. (The former owner of a horse, with no ties to the present owner, will usually tell you the truth about the horse.)

COMMON BREEDS OF HORSES

Breed	Normal usage	Comments
Arabian	Pleasure (E & W) Park Stock horse Trail riding Endurance riding Parade Pleasure Driving	Exceptional endurance Foundation breed of most other purebread breeds
Appaloosa	Pleasure (E & W) Stock horse Trail riding Endurance riding Reining Cutting Parade Gymkhana, timed events Pleasure driving Roping Polo	A color breed Not purebred
American Saddle Horse	3- and 5-gaited Pleasure Fine harness Pleasure driving	Mostly used in show ring and on bridle path
Morgan	Park Stock horse Trail riding Pleasure driving Roadster	Descended from one foundation sire (Justin Morgan)
Paint	Pleasure Stock horse Reining, roping Cutting Racing (short) Parade Trail riding Polo	Color breed of Quarter Horse conformation. Two basic color patterns: Overo and Tobiano
Pinto	Used more extensively under English saddle Polo Parade	Color breed Conformation may be characteristic of most of the English type breeds Overo and Tobiano
Palomino	Pleasure (E & W) Stock horse Parade	Color breed Many double registered, especially Quarter Horses
Quarter Horse	Pleasure Stock horse Reining, roping Cutting Racing (short distance) Trail riding Gymkhana Timed events Polo Hunters Dressage	Largest in registration of all breeds. Due to its versatility and temperament, a very popular breed

Breed	Normal usage	Comments
Standardbred	Harness Racing—trotters and pacers Roadsters	Used almost exclusively as racing and driving horses
Thoroughbred	Racing Hunters Jumpers Polo Stock horses	Used extensively to add quality, speed, and endurance to many of the other breeds
Tennessee Walking	Show ring Pleasure	A characteristic running walk; when developed to the extreme is used mostly in the show ring
Connemara	Medium-sized adult's and children's mounts Hunters Jumpers	Noted for their temperament, hardiness, and jumping ability
Pony of Americas	Children's mount Pleasure (E & W) Timed events	46 to 52 inches high and have Appaloosa markings
Hackney Pony	Harness show ponies	Extreme flexion in the knees and hocks. Probably the showiest of all at the trot
Shetland Pony	Children's mount Pleasure driving Harness show Pony (American type)	Old Island type short and blocky, and new American type modeled after the American saddle horse
Welsh Pony	Children's mount Roadster and racing ponies Hunter ponies	A very hardy pony

Note: E & W refers to English and Western saddles.

Space and Fencing

In most communities, keeping horses in heavily populated areas has precipitated the inclusion of animal ordinances in zoning regulations. Laws governing cruelty to animals, nuisances, health, and safety are usually already on the books. Be sure to check all governmental requirements.

Pastures should be high in fertility of soil and density of desirable forage species in order to provide enough feed for the horse. Your county Extension agent can give sound advice.

If the horse is to be confined with limited riding, about 800 square feet should be provided for a minimum exercise area. Longeing the horse for half an hour a day can be done with very limited space. (Longeing is having the horse circle you on a long 25- to 30-foot line.)

Fencing for horses usually involves more expense due to the temperament of the horse. Barbed wire, one of the most inexpensive fencing materials, is the least desirable for horses. The horse is easily frightened and in small pastures is apt to try to run through barbed wire, which frequently causes grave injury.

Suitable types of fences include:

- Wood—plank, board, split rail, rail, and buck and rail. Wood is the safest fencing for horses, but expensive to construct and maintain.
- Metal—woven wire, chain link, welded pipe, cable, barbless wire, and plain wire.
- Electric—Use only an approved, safe system. Follow directions for installation. Horses must be trained to an electric fence, which can be used to divide pastures for rotational grazing.

All wire or metal fences should be grounded at least every 800 to 900 running feet to a metal rod driven into the ground to permanent moisture depth.

Fence Posts

Locust, red cedar, and osage orange are the best wooden posts available. Railroad ties in good condition have long life, are strong, and hold staples well. Pressure-creosoted, discarded telephone poles may also be used.

All other wood posts—such as white cedar, oak, hemlock, hickory, pine, spruce, birch, beech, and ash—should be treated with a wood preservative such as creosote or pentachlorophenol. Note that with metal posts, a T-shaped driven post is stronger than U- or V-shaped posts. Concrete posts have long life but are expensive.

Managing Pasture

Many horse pastures are no more than exercise lots. The quantity and quality of feed on these run-down pastures won't meet the horse's nutritional needs. Pleasure horses usually graze limited areas. Therefore, try to balance the needs of the horse with the needs of the pasture.

To improve soil fertility, apply lime and fertilizer. Eliminate brush and weeds to provide more sunlight, nutrients, and water for the desirable pasture plants. Drainage may be required to improve the growth environment. Reseeding to change the plant composition will increase forage productivity. Large pastures divided into several smaller pastures and rotated during the grazing season provide more forage per acre.

Feed Requirements

To satisfy the horse's nutrient requirements, you need to provide: a balanced ration in amounts large enough for growth (if immature); energy (for body temperature regulation, vital body functions and work, and the repair of worn-out body tissues); development of the young (pregnant mare); and production of milk (lactating mare).

A balanced ration must contain the correct amounts of:

PROTEIN. Vegetable seed derivatives such as soybean oil meal furnish more of the essential amino acids.

ENERGY. Energy is derived primarily from the carbohydrate and fat portion of the ration.

VITAMINS. A and D are the most important. Horses kept on pasture or with access to the sun normally assimilate enough vitamin D. Stabled animals should be supplemented. Vitamin A is derived primarily from green leafy roughages (hay) in the form of carotene.

MINERALS. Calcium and phosphorus in proper balance (most agree on a ratio of 1.1:1), and salt in proper amounts are vital.

Clean water is essential for digestion and should be provided and accessible at all times (except for hot horses or in other cases where intake must be controlled). A horse consumes an average of 10 to 12 gallons per day.

Each bag of commercial feed has a tag attached. The tag lists the percentage of crude protein, fiber, and fat, and the feed ingredients.

Pelleted feeds are popular with some horse owners because of ease of storage, less dust, less manure, and less waste. The big disadvantage is that most horses develop a wood-chewing habit if pellets are fed without some hay.

Cut good hay at the proper stage of maturity. It should have a bright green color, an abundance of firmly attached leaves, and a sweet and pleasant smell.

Tips on Feeding

Savings on feed costs can be realized by prudent buying, proper care of the horse's teeth, elimination of parasites, feeding according to need, reduction of waste, and clean water in sufficient amounts.

Follow a well-planned feeding program. Since the horse is a creature of habit, a poor feeding schedule can establish bad habits in the stall. Crowding, kicking, pawing, and biting are some of the most common problems. Giving a horse "treats" by hand can establish a habit of nipping or biting.

Several feed management practices that should be followed are:

- Feed each horse as an individual and know its weight and age. Feed at regular times. Use clean feed boxes.
- Horses doing hard work need more grain and less hay. Split grain ration into three or four feedings per day with the biggest portion at night. Split hay into a quarter in the morning, a quarter at noon, and half at night.
- Use a good-quality, nutritionally balanced feed. Never feed moldy, dusty, or frozen feed. Feed by weight, not by measure.
- Have salt and mineral mix available to the horse.
- Check teeth regularly.
- Check the three Hs—hair, hide, and hoof—for condition. They are good indicators of digestive problems.
- Check the consistency and odor of droppings.
- Control parasites, internal and external.
- Don't make abrupt changes in feed. This also applies to newly cut hay.
- Don't feed large amounts of garden produce to horses.
- When first turning your horses out to pasture in the spring, be sure to give them a full feed of hay prior to turning out and then some hay each day for the first week. The first few times, turn out for only two to three hours until they become accustomed to the new lush growth.
- Don't throw clippings from yew bushes in with horses. Remove broken limbs of wild cherry trees that contain wilted leaves. Both are poisonous.

RULE OF THUMB FEED REQUIREMENTS FOR A MATURE RIDING HORSE		
Use	Commercial mix grain 10–12% protein	Hay
Light 1–3 hrs/day	⅓ to ½ lb. per 100 lbs. body weight	1¼ to 1½ lbs. per 100 lbs. body weight
Medium 3–5 hrs/day	¾ to 1 lb. per 100 lbs. body weight	1 to 1¼ lbs. per 100 lbs. body weight
Heavy 5–8 hrs/day	1 to 1½ lbs. per 100 lbs. body weight	1 to 1¼ lbs. per 100 lbs. body weight

Stables and Equipment

In the Northeast and areas plagued by rapid changes of weather, it is important to provide shelter for horses. A three-sided shed open to the south or southeast and protected from the prevailing winds, well bedded and free from drafts, will provide all the shelter needed. Provide horses kept outside with a slightly higher plane of nutrition, to meet the greater requirements for body heat.

Horses maintained in show condition either are blanketed and housed in enclosed stables, or heat is provided in the stable. The amount of heating will dictate the amount of ventilation and insulation that must be incorporated in order to control condensation. Doors, windows, louvers, and cracks will usually provide enough ventilation in an unheated stable, but you should avoid drafts.

An ambient temperature of 32 degrees Fahrenheit will prevent freezing of water pipes and also be very comfortable for the horse. Stale air, condensation, dust, and drafts are all harmful.

A good stable design with attractive fencing does much to make a horse acceptable in a suburban area. Incorporated into the design should be a rodentproof feed room, hay and bedding storage, a moistureproof tack room, and provision for manure storage.

Box stalls or straight stalls may be used for horses or ponies. Provide a 5-gallon water bucket. Cement is not recommended as a floor material. It is slippery, cold, and hard on the horse's feet and legs. Planking is used in many areas, but it rots and has to be replaced periodically, creates odors, and provides an excellent environment for mice and rats.

The preference of most horse owners is tamped clay free of rocks. You might want sandy loam free of rocks or sand, but during the winter, sand makes a cold and damp bed for the horse. Then too, horses may develop sand colic through ingesting sand while eating spilled feed.

Many types of bedding are available: straw, sawdust, shavings, peat moss, peanut shells, even leaves and pine needles. Some are more absorbent than others and the price range is wide. A diligent sorting of soiled bedding can cut down costs and the amount of wastes to be disposed of.

Place one window in each stall to provide light and ventilation. Have one electric lighting fixture over every other stall, and lighting for tack and feed room and alleys. All fixtures should be covered by dustproof shields; wiring and switches should be out of reach of the horses or protected.

The feed room should be horse- and rodentproof, and provide for tool storage. Completely enclose the tack room. If heated, it should be completely insulated. Keep it clean and free of dust and moisture, which can cause rapid deterioration of leather. The tack room should include blanket, saddle, and bridle racks, a first-aid cabinet, a saddle cleaning stand, and equipment storage. Boxes can be provided near stalls for grooming equipment.

Training or exercise rings do not require as much space as show rings. Normally a 40-by-80-foot ring will suffice. Careful planning can create extensive and interesting trails in small wooded areas. Neighbors with horses and land can develop connecting trails for mutual use. Horse owners with only stabling facilities and no nearby trails or rings available will probably have to invest in a horse trailer. Most owners eventually do.

Tack requirements consist of a halter and lead rope, bridle and saddle (English or Western), saddle pad or blanket. The rider's personal attire will depend on the style of riding.

Health Care

Practice preventive medicine. Provide adequate shelter, safe fencing, a balanced ration, good dental and foot care, a sanitary environment, control of parasites, and immunization against diseases.

The second line of defense, and the most important, is to pick a good veterinarian. A veterinarian who specializes in equine medicine is probably the best choice; second choice is one who has a large-animal practice. Check with local horse owners for their recommendations.

The vet will recommend immunizations for your area. Ask for and follow the vet's parasite control program.

Learn to recognize the seriousness of wounds, know how to treat them, and when to call the vet. The vet can instruct you in follow-up nursing care. With experience and study, many problems can be taken care of by the owner.

A good farrier is necessary for the health of the horse's feet. Most farriers can recognize diseased feet. Horses should have their feet trimmed regularly if kept unshod. If shod, the shoes should be reset or replaced every six to eight weeks.

Routinely clean out the horse's feet. This, along with daily removal of soiled bedding and manure, and replacement with fresh bedding, helps prevent diseased feet.

Vacation Farms

Times are very hard for farmers. Even if your property is just a few acres, consider the value of agritourism as an income stream. People are eager to experience farm life, and you are eager to keep your farm running in the black.

If you looked around the countryside at vacation farms, you would be struck by their great diversity. Some vacation farms have animals. Some have guest rooms in the farmhouse, others have cottages. Some offer fully prepared meals, others provide cooking facilities. Some have swimming pools and tennis courts, others simply offer the "ole swimming hole." The rural environment with open space and clean air is common to all.

Vacation farms continue to be popular in the twenty-first century. Roughly 52,000 U.S. farms or about 3 percent of all farms, earned income with farm-based recreation, bringing in about $955 million in 2004. The USDA believes that agritourism could become bigger. What works against a hotel—quirkiness, unpredictability—works to your advantage as a vacation farm. Some run like bed-and-breakfast inns, while others offer activities such as swimming, the opportunity to help with farm chores, and hayrides. The typical farm or ranch vacationer can be described as highly educated, and holding a professional or managerial position with relatively high earnings. Most typically, this vacationer also has a larger-than-average family.

Generally, farm vacationers do not seek a highly organized program of recreational activities, but they desire to participate in a variety of activities. Horseback riding, hiking, and swimming all rank high on the list of desired activities. A single activity, like observing farm animals, is not enough. Urbanites seek to place themselves temporarily in the clean life-style of rural America.

A vacation farm operator and his family must like working with people, often as much as eighteen hours a day. This calls for time, patience, and understanding. You will lose a certain amount of privacy, but a better understanding of people is almost sure to rub off on the farm family.

Farm vacationers need private quarters—their own dining room, lounging area, and toilet facilities. Cleanliness is an absolute necessity, not only to attract and keep guests, but also to meet health department requirements. Accommodations should be cozy; there is nothing like a crackling fire on a cool or damp night.

Water is a key word in more ways than one. Provide guests with an abundant supply of pure water under pressure, between 100 and 150 gallons a person per day. Additionally, clean water for recreation is highly desirable. Swimming is a top recreational activity during summer months. A pool is often preferred over a pond, and quality is easier to control.

Good, old-fashioned, home-cooked meals loaded with variety and imagination keep folks happy. Guests are on a holiday schedule, so meal hours must be flexible.

Clientele Goals

Think carefully about your clientele. You are in the unique position of getting labor and acting as innkeeper. Do you want to cater to young families, or older couples without children? Although some operators attract both groups, this can lead to problems. Older couples like a quiet setting, while children like lots of activity. A vacation farm catering to young families has a limited season, tied to summer vacations. On most farms, this coincides with the busiest harvesting schedule. Carefully evaluate the potential conflict between both activities.

Unless you plan to run a summer camp, it's sensible not to accept children without their parents. But many farms operate as something closer to a bed-and-breakfast for families, finding that so many tourists ache to experience farm life. Your state might have a farm vacation organization that offers ideas and a way to advertise your place.

Young adults are equally interested, and they don't mind working. An organization called World Wide Opportunities on Organic Farms helps those in their twenties (usually) find work on organic farms in exchange for room and board. WWOOF, as it is called, lists farms looking for workers. Potential volunteers contact farmers directly. See www.wwoof.org. What about the chores? It seems that many successful vacation farms find that even when cleaning the barn or picking up eggs is just optional, many of the guests leap to do it. (Can you hear your overworked farming ancestors turning in their graves from amazement?)

Legal liability for accidents or injuries represents another basic consideration. Some activities pose greater risk than others. Insurance coverage varies in cost, depending on which activities you specifically list in your policy. Discuss every activity you plan to offer with your insurance agent. Incorporation becomes an important consideration, too, as a technique that limits your liability.

Fringe benefits accrue as you establish your vacation farm. The tennis court or swimming pool you build for the vacationers are great places for you to relax during the off-peak times.

How to Charge

Nationally, vacation farms are relatively few in number and diverse in character. For this reason there is no established food and lodging price range that can be considered appropriate for all. However, a few guidelines do apply. Business should be attracted on the basis of recreational opportunity, not low cost. Some budgeting of potential income and expense at several different prices will be helpful. Income must cover all costs and return a reasonable profit for time spent.

Income on the vacation farm is not restricted to food and lodging charges. Farm products like maple syrup can be sold directly to your guests at retail prices, as can fishing tackle and bait to the fishermen. Boat rentals will also be profitable.

The camper needs a variety of services such as ice, firewood, bottled gas, and food supplies. It is not uncommon for campers to spend more money on services than on camping fees, but don't nickel-and-dime your guests excessively. Vacationers are willing to pay a reasonable total price but object to a constant bombardment of small charges.

Several features of a vacation farm give it greater appeal than standard resorts. Prices are usually lower because the farm operator has less invested than the traditional resort operator. Farms are less crowded than resorts. And country living has an attractive new image, especially among young people.

The RV Market

The recreation vehicle (camper, trailer, or motor home), so popular today, must be considered in the total picture. Currently, recreation vehicle (RV) manufacturers are shipping over half a million units annually to dealers. RV owners seek a place to park their unit where electricity and water are available, and clean toilet facilities conveniently located. Catering to this group is easy for the farm operator—no sleeping accommodations are needed. While the potential for vacation farms looks good, it will never be fully realized without a significant marketing effort. Consumers are bombarded by advertising and promotional programs from other recreation suppliers looking for a share of the market.

Vacation farm operators might well look to their old friend, the cooperative, for help. As often demonstrated, a group of farmers advertising together can achieve results that would be impossible for one alone.

Don't make any decisions about this business without first visiting at least one ongoing vacation farm. Go as a regular guest—a great deal more can be learned. While you rub elbows with other guests, find out their likes and dislikes. Talk to the management about problems and opportunities, and stay objective during this evaluation process.

Take part in recreational activities offered, both familiar and unfamiliar. This can be fun and you may uncover useful ideas. Keep notes on your thoughts and impressions and take plenty of pictures. These can help you evaluate your experience when you get back home.

Before you start your business, visit the state health department or licensing authority. Find out about state standards. If your planned facilities are substandard, how much will it cost to bring them up to the mark?

Gather the entire family and discuss, in frank terms, potential impact of the venture on family life. Is everyone ready for this change? Carefully weigh all benefits against all costs, both monetary and nonmonetary. Take a vote. If the "ayes" have it, welcome aboard. Good luck!

PART 6

Property

Disposing of Property

What if you decide that your experiment in rural living must come to an end? Or you begin to think about what will happen after you are gone? You are ready to sell some or all of your property, or to create a will that will establish what happens to your land in the future. Turn to professionals for guidance.

If you own a large piece of undeveloped or active farm acreage, consider the options for preserving it. There are ways to preserve rural tracts as farmland or open space while getting a fair price (see the section that follows, "Transferring Land with Preservation in Mind"). In today's climate, rural land is becoming ever more precious, and as a seller you can control what will happen to it.

The Basics of Selling Property

First, decide whether to employ a real estate broker or sell the property without one. Decide which assets (such as farm equipment) you want to include, and your asking price. It is on this latter point that a real estate broker can be of great assistance. In addition, services of an appraiser may be desirable to help determine the property's current value.

If you enlist the aid of a broker, agree on the terms of the broker's employment, and spell it out in a listing contract. Standard contracts have room for unique details of your arrangement. Basic types of listing contracts include open listing, net listing, exclusive listing, and exclusive right to sell.

The open listing may be either an oral or written contract. It is a simple agreement in which the seller agrees to pay a stated commission if the broker obtains a buyer to sign a purchase contract agreeable to the seller. This does not preclude the owner from making the sale; nor does it preclude contracts with other brokers.

Although there are certain advantages to the seller, this type of contract is not favored by most brokers and may have disadvantages for the seller as well. The broker may not be interested, knowing other brokers will also be authorized to obtain buyers for the property. And a broker who does sign up is less likely to actively promote and advertise the property.

In a net listing contract, you set a low price below which the broker can't list your property. Generally the real estate broker is authorized to add the commission or fee over and above this base amount. Many owners like this arrangement, but most brokers do not. A more common type of listing contract is the exclusive listing, preferred by most real estate people. One broker is appointed to act as agent for the seller for a set time—often for three to six months.

A similar arrangement is the exclusive right to sell. But here the broker is entitled to a commission if the property is sold at any time during the contract term, even if you, the owner, arrange the sale. As a practical matter, most brokers prefer this arrangement. However, many sellers often have contacted potential buyers on their own before the property is listed. So it is fairly common practice to modify the exclusive right-to-sell contract by providing that the broker gets no commission if a sale is concluded with a buyer that you previously contacted.

With your broker, specify the sale price, terms of the sale, and whether you will include personal property (such as equipment). Understand how the commission will be calculated. Generally the selling price drives the amount of the broker's commission, which can vary from 5 to 10 percent, depending on the type of property.

Other provisos of the listing contract may specify the amount of "down payment" or "earnest money" expected from a potential buyer, and what is to be done with this money until final transfer of the property.

In many areas, brokers offer a multiple listing service. With this practice, many brokers can be authorized to sell the same property at the same time. Many multi-listings provide for sharing the commission between the real estate office obtaining the listing and the office that arranges the sale. This benefits you because it advertises your property far and wide.

Negotiating the Contract

Once an interested buyer comes forward, the real estate broker negotiates a sales contract—including price, description of the property, financing arrangements, title search, personal property to be sold, and agreements on taxes—between you and the buyer. In most areas, brokers are authorized to prepare the sales contract. It may also be prepared by a lawyer, and in some states local law requires a lawyer.

Many contracts require a survey and various inspections. Once the parties reach an agreement, the buyer usually makes a deposit, most often to the real estate agent, who puts it into a bank escrow account. A deposit of 10 percent of the purchase price is ordinarily required. This will be applied toward the total purchase price.

Title Examination

Most real estate sales contracts provide for the seller to deliver "marketable title" to the property. That means title can be transferred free of reasonable doubt as to its validity, to avoid lawsuits later. Lawyers or other professionals do this. If no title defects appear, the sale can be concluded. If the lawyer unearths problems with the title, you might be given time to correct any defects (such as an unclear boundaries). If the questions can't be resolved, the buyer may be allowed to withdraw.

Title examinations require a thorough search of courthouse records in the county where the property is located. In some states, professional abstractors do the search, and they furnish a certified summary called the "abstract of title." Your lawyer can examine this document. In other states the lawyer handles the search directly.

The title searcher examines every deed of sale for your land, going back sixty or more years. In some areas the chain of title may be traced back to the original government land patent of a century or more ago. The title search also calls for a look at other public records that may affect the title. After all records have been checked, a title report is submitted to the potential buyer. This report gives full information regarding title to the property. An attorney may be asked to give an opinion on "marketability" of the title.

In many areas, the buyer buys title insurance, which insures against defects that would normally appear in the records. Normally the buyer pays the cost. However, the sales contract may provide that the cost be paid by you, the seller, so it's a good idea to ask the broker about this beforehand.

During the period from the time of the sales contract until closing date of the transaction, the buyer will secure a loan, if necessary. Certain types of financing arrangements require the seller to be directly involved. For example, you may arrange for the buyer to give you a promissory note and pay you directly in installments, or the buyer may assume a loan that you took out.

While the buyer usually pays the bank "points"—the lender's charge above the interest rate—on some loans the seller may be required to bear this expense. One point generally amounts to 1 percent of the mortgage amount, and because the number of points varies, this expense can be thousands of dollars, so pay attention.

Transferring Land with Preservation in Mind

If you own several country acres, selling this land could mean it will become a housing development. Do you want this to happen? If not, there are ways to sell the property while stipulating that it remain as it is. Farmland and open space continue to disappear in this country. The reality is that farmland is worth more, per acre, to housing developers than it is in other uses. As more farmland becomes developed, new legal mechanisms to save it have emerged.

Some states offer farmland preservation programs in which you sell the development rights to the state government while maintaining ownership of the land. Contact your state agriculture department. You also might consider selling the land to a land trust or other private organization that buys land to preserve it.

If you want to remain where you are but pay less in taxes, you can change the legal designation of your land by granting a conservation easement to a land preservation group. In that case, you would continue to own the land, but a legal document (sometimes quite lengthy) would stipulate exactly what activities can and cannot take place on it. You might retain the rights to cut timber or hay or graze animals, for example, while relinquishing the rights to build on the land or to sell it to someone else who would.

Identify Your Values and Needs

When planning the future of your land, think of a three-legged stool with the land in the middle. How much of the land is good farmland, or hillside pasture, or a woodlot? Then think about the three major questions.

First, what does your family want to do with the land in the future? Second, what are the family's values where land conservation is concerned? Third, what are the family's financial needs?

Depending on how much land you own, there are various options for protecting a good portion of the land, while selling off a few lots for houses. You might protect the land that is good for farming, while setting aside some house lots that are guaranteed to have a pastoral view, making them more valuable on the open market and providing you with some income.

In 1980, the American Farmland Trust was formed. It has protected more than one million acres of farmland since then. It helps owners of rural land to negotiate conservation easements. It also encourages good planning by land-use boards and distributes information.

For information about programs to preserve farmland or other open land, contact the following organizations:

American Farmland Trust, 1920 N Street, NW, Suite 400, Washington, DC 0036. Call (202) 659-5170. The Web site is www.farmland.org.

Land Trust Alliance, 1319 F Street, NW, Suite 501, Washington, DC 20004. Call (202) 638-4725.

Trust for Public Land, 116 New Montgomery Street, 4th Floor, San Francisco, CA 94105. Call (415) 495-4014.

Providing for Your Heirs— Nonsale Property

Leaving Your Land to Heirs

If you want to turn over your land without selling, you are either passing it on to relatives, or trying to preserve it for nature.

Basically, these transfers fall into two general classes: lifetime gifts and testamentary bequests. Motivations for each may differ, partly depending on legal and tax considerations. These motives are better understood by considering your objectives in planning your estate. You must consider the land's ability to provide you with some income during retirement. You might want to turn over some of the land to your children or other loved ones. Finally, you want to save those heirs from having to pay the expenses and taxes associated with transferring property to the next generation. Minimizing these costs maximizes the value of the property transferred to heirs.

Get Professional Help to Understand Tax Codes

Work with a competent estate planner to understand all of the implications in transferring your land, and to minimize transfer expenses, such as gift and estate taxes and probate court expenses. Generally speaking, your goal is to minimize transfer costs, but you must also consider what you want to happen to the land.

A crucial point to consider is how your land is designated. Remember that if your land is officially designated as farmland, it will not cost your heirs as much as it would if it is valued as land that can be developed (which would be considered more valuable) under the federal tax code. If your farm business meets certain qualifications, family farms can remain so for the next generations, without costing the heirs so much that they can't afford to keep going.

Most states have either an inheritance or an estate tax. An inheritance tax is a tax on the right to receive property, and generally is based on the value of the property. Many states provide tax-free exemptions for some specific value of inheritance.

An estate tax is a tax on the right to transfer property. Therefore, it is based on the total value of property owned (or controlled) by the decedent. In general, the amount of taxes collected under state inheritance and estate taxes are not greatly different. You must consider federal taxes as well. If you make a lifetime gift of your land, you pay a tax on that gift. If you leave the land to someone in your will, the recipient pays a tax.

Planning Tools

Wills, trusts, joint ownerships, gifts, sales, family annuities, and other arrangements may be used to achieve estate planning objectives, depending on family circumstances.

WILLS. Basically, anyone owning property or having children needs a will. Furthermore, each spouse should have a will, particularly if there are minor children, because generally only the longer-surviving spouse may legally name a guardian for them.

Obviously, wills may be used to achieve the desired property distribution. But they also may minimize death taxes and probate expenses; keep the property out of the hands of children until they are mature enough to manage it; name guardians for children and property; name who will handle your estate; and preserve the family business.

Consult a knowledgeable lawyer to explore ways to transfer your property by establishing co-ownership, joint tenancy, or community property.

Holding property with family members in joint tenancy is sometimes called a "poor man's will" because it achieves the desired property distribution. (It does not provide the other functions of a will.) However, joint tenancies can cause adverse tax consequences if an estate is subject to federal gift and estate taxes.

Types of Trusts

A trust involves a nonsale transfer of property to a trustee, who invests and manages it for the benefit of the beneficiaries. These beneficiaries can be anyone selected by the trust creator, called the grantor. The trustee may be a family member, someone from your bank, or a trust lawyer. Who takes that role depends on the kind of property in the trust and the expertise of alternative trustees in managing different types of property.

Some trustees may be adept at managing a diversified stock portfolio; others are skilled in farm management.

There are two basic kinds of trusts:

A **TESTAMENTARY TRUST** is created in your will. For example, you might set up a trust to benefit minor children, distributing income to them for support and maintenance but retaining title until they are mature enough to wisely handle ownership and management decisions.

LIFETIME TRUSTS may be either revocable or irrevocable. Under a revocable trust you, the grantor, retain the power to take the property back. If the trust still exists when you die, the trust property need not go through the probate court, and it may be distributed outside the public view.

The fact that trust distributions are not a matter of public record (while probate matters are) is thought by some to be an advantage. However, the existence of the power to revoke is sufficient to include the value of trust property in the grantor's estate when computing federal estate taxes due.

An irrevocable trust is one that you, as the grantor, cannot revoke. Since it constitutes a completed gift, its creation may be a taxable gift for federal gift tax purposes.

Trusts cost money to create and operate (trustees are paid for their services). But they may minimize or avoid death taxes and probate expenses, relieve others of ownership and management obligations, assure high-level management, minimize income taxes, and achieve desired property distribution at your preferred point in time.

How to Get Help

If you want your garden soil tested or if you have questions about gardening, livestock, insects, plants or animal diseases, or other questions about living on a few acres, contact your county Cooperative Extension Service

Some pesticides are for restricted use only, and you must be certified before you can purchase or use them. Your county Extension agent can tell you which pesticides are restricted and can also suggest alternative chemicals or methods for controlling pests. In addition, your agent can tell you how to become certified if that is necessary.

County Extension offices are usually located at the county seat in the post office, the courthouse, or in a building shared by other USDA agencies.

Cooperative Extension offices are usually operated through state universities or housed in buildings shared by other USDA agencies. The U.S. Natural Resource Conservation Service, working through the local soil and water conservation district, can help you find out what your land and water resources are like and how to improve them in order to grow better crops, support more wildlife, boost the supply and quality of water, and produce more income.

The Farmers Home Administration makes loans to farm and non-farm families alike for homeownership and improvement in rural areas.

Young people of student age—minors—may get loans on their own signature for income-producing projects carried out as part of a youth organization (4-H, Future Farmers of America) or school programs, such as a garden cultivated for purposes of selling the produce.

Help from the Web

American Small Farm magazine, http://www.smallfarm.com

California Certified Organic Farmers, http://www.ccof.org

Midwest Organic and Sustainable Education Service, http://www.mosesorganic.org

New England Small Farm Institute, http://www.smallfarm.org

Northeast Organic Farming Association, http://www.nofa.org

Small Farm Digest. For information: http://www.nifa.usda.gov/newsroom/newsletters/smallfarmdigest/sfd.html

Southern Sustainable Agriculture Working group, http://www.ssawg.org

USDA National Institute of Food and Agriculture, http://www.nifa.usda.gov/nea/ag_systems/in_focus/small_farms.html

Notes

Foreword

For demographic trends, we consulted Kenneth M. Johnson and Calvin Beale, *The Rural Rebound: Recent Nonmetropolitan Demographic Trends in the United States*. Loyola University, Chicago, 2003; and the U.S. Census Bureau.

Part 3

LAND IMPROVEMENTS—WHAT YOU NEED TO KNOW

For the cost of barbed wire and woven wire fencing, we consulted Iowa State University Continuing Education and Communication Services publication, "Estimated Costs for Livestock Fencing," 1999.

For the cost of electric fences, we quoted a price from Kencove Fence Supplies of Blairsville, Pennsylvania.

For the cost of gravel, we consulted several public and private sources and quoted a price from Johnson's Nursery and Landscaping Inc. of Berryville, Arkansas.

WATER AND WASTE DISPOSAL

For the minimum daily water requirements, the editor contacted Peter Gleick, president of the Pacific Institute for Studies in Development, Environment, and Security, and author of *The World's Water: The Biennial Report on Freshwater Resources, 2000–2001*.

For a wealth of information on small-wind energy systems on farms, see the American Wind Energy Association website, http://www.awea.org/smallwind/smsyslst.html.

Part 4

FIVE YEARS ON FIVE ACRES

The source for cost estimates on raising a pig came from Kelly Klober, *Storey's Guide to Raising Pigs,* 1997.

Part 5

VEGETABLES

For information on starting or selling through CSAs or farmers' markets, we consulted the USDA's Agricultural Marketing Service, which maintains an extensive collection of advice at www.ams.usda.gov. The USDA's Alternative Farming Systems Information Center's National Agricultural Library in Bethesda, Maryland, makes available many publications on CSAs. For a list, see http://afsic.nal.usda.gov.

The CSA resource is a 2009 USDA publication produced by the Alternative Farming Systems Information Center of the National Agricultural Library. For information, see Web site: http://afsic.nal.usda.gov/.

BEEKEEPING

For information about the Africanized bees, we consulted a USDA report of the department's Agricultural Research Service, "Answers to the Puzzling Distribution of Africanized Bees in the United States," by J. D. Villa et al., 2002. See http://msa.ars.usda.gov/la/btn/hbb/reprints.htm.

For information about tracheal mites, we consulted, "A Survey of Tracheal Mite Resistance Levels in U.S. Commercial Queen Breeder Colonies," by Robert G. Danka and Joseph D. Villa, Agricultural Research Service, USDA. See http://msa.ars.usda.gov/la/btn/hbb/reprints.htm.

CHRISTMAS TREES

For updated information about Christmas tree growing, we consulted the University of Illinois Cooperative Extension. See http://www.urbanaext.uiuc.edu/trees/treefacts.html.

GROWING NUT TREES

For latest growing figures for English walnuts, we consulted the Agricultural Marketing Resource Center, Iowa State University, Ames, IA. See http://www.agmrc.org.

KEEPING CHICKENS AND OTHER FOWL

For the cost of started pullets, we consulted Virginia Cooperate Extension Service, "Small Flock Fact Sheet No. 30." See http://www.ext.vt.edu.

Raising Rabbits

For updated information about the number of rabbits raised in the United States, we consulted the Mississippi State University Extension Service. See http://www.msstate.edu.

The net profit estimate comes from "Agricultural Alternatives: Rabbit Production," a pamphlet of the Pennsylvania State University College of Agricultural Sciences Cooperative Extension. See http://agalternatives.aers.psu.edu.

Raising Sheep Part-Time

For updated data about marketing sheep, we consulted the Economic Research Service of the USDA. See http://www.ers.usda.gov/Briefing/Sheep/.

Part 6

Transferring Land with Preservation in Mind

We consulted the following resources about land preservation programs:

We credit Frederick B. Gahagan, a lawyer specializing in land conservation in Old Lyme, Connecticut, with the "three-legged stool" approach to preserving rural land.

"Purchase of Development Rights," by Joe Daubenmire and Thomas W. Blaine. Ohio State University Cooperative Extension Fact Sheet CDFS-1263-98. Available online at ohioline.osu.edu/cd-fact/1263.html. Or contact OSU Cooperative Extension Community Development, 700 Ackerman Road, Columbus, OH 43502-1578.

"Conservation Easements," by Peggy Schear and Thomas W. Blaine. Ohio State University Cooperative Extension Fact Sheet CDFS 1261-98. Available online at http://ohioline.osu.edu/cd-fact/1261.html. Or contact OSU Cooperative Extension Community Development, 700 Ackerman Road, Columbus, OH 43502-1578.

Further Reading

Classics

Editor's Note: The book you are holding is the updated version of one of the true classic how-to books for small-scale homesteaders. Here are some others.

Five Acres and Independence, by M. G. Kains. Basic information from an expert writing in 1940.

The Good Life: Helen and Scott Nearing's Sixty Years of Independent Living. From the 1930s to the 1990s, these former New York City academics carved out a self-sufficient farm that became a landmark. Schocken Books, 1990.

A Pattern Language, by Christopher Alexander et al. This classic book describes ways to make your house and other buildings in sync with the natural environment. Oxford University Press, 1977.

Advice

Basic Country Skills, by John and Martha Storey. Illustrated guide to becoming self-reliant. Garden Way Books.

"Conservation Easements," by Peggy Schlar and Thomas W. Blaine. Ohio State University Fact Sheet CDFS-1261-98. (Available at http://ohioline.osu.edu or by contacting Ohio State University Community Development, 700 Ackerman Road, Columbus, OH 43202-1578.)

Considerations for Agritourism Development, by Diane Kuehn et al. New York Sea Grant, State University of New York at Oswego, 1998 and 2000. Cornell University Web site: http://ccc.cornell.edu/seagrant/tourism/www.agrifs.pdf.

The Encyclopedia of Country Living, by Carla Emery.

Farm & Ranch Recreation Handbook, by Susan J. Rottman and Jeff Powell. 2002. Cheyenne, Wyoming: RLS International. Web site: http://uwadmnweb.uwyo .edu/RanchRecr/handbook/table of contents.htm.

Promoting Tourism in Rural America, compiled by Liam R. Kennedy, Clarion University of Pennsylvania. Rural Information Center Publication Series, no. 60, revised edition, 1998. Rural Information Center, National Agricultural Library, USDA/ARS, Beltsville, MD 20705, (800) 633-7701.

"Purchase of Development Rights," by Joe Daubenmire and Thomas W. Blaine. Ohio State University Fact Sheet CDFS-1263-98. (Available at http://ohioline .osu.edu or by contacting Ohio State University Community Development, 700 Ackerman Road, Columbus, OH 43202-1578.)

The Rodale Book of Composting, revised edition, edited by Grace Gershuny and Deborah Martin. Rodale Books, 1992.

Stories Across America: Opportunities for Rural Tourism, by Suzanne Dane et al. Contact the Rural Information Center, National Agricultural Library, USDA/ ARS, 10301 Baltimore Ave., Room 304, Beltsville, MD 20705, (800) 633-7701.

Contact the MidWest Plan Service, 122 Davidson Hall, Iowa State University, Ames, IA, 50011, (800) 562-3618, www.mwpshq.org, if you are interested in any of the following titles:

- *The Beef Cattle Handbook,* by the Beef Cattle Resource Committee of the North Central Land Grant Universities. June 1999.
- *Enhancing Wildlife Habitats: A Practical Guide for Forest Landowners.* 172 pages, 1993.
- *Fire Control in Livestock Buildings.* How to build to avoid disaster. 18 pages, 1989.
- *Greenhouses for Homeowners and Gardeners.* 2000.
- *A Guide to Logging Aesthetics: Practical Tips for Loggers, Foresters, and Landowners.*
- *Home and Yard Improvements Handbook.* Classic first edition, 100 pages, 1971.
- *Horse Housing and Equipment Handbook.* The classic 1971 60-page booklet.
- *House Planning Handbook.* An 81-page classic on matching your life to your house. Second edition, 1988.
- *Onsite Domestic Sewage Disposal Handbook.* 40 pages, 1982.
- *Produce Handling for Direct Marketing.* How to store, move, and display produce. 1992, 29 pages.
- *Sheep Housing and Equipment Handbook.* Fourth edition, 1994.
- *The Small Dairy Resource Book,* by Vicki Dunaway.

Contributors to the USDA 1978 Yearbook of Agriculture

Living on an Acre is the updated version of *Living on a Few Acres,* the United States Department of Agriculture's 1978 Yearbook of Agriculture. We acknowledge below all of the original authors, whose names appear in the order of their appearance in this edition:

Part 1

Bob Bergland, former USDA Secretary of Agriculture
A. Gene Nelson and Tom Gentle
Manning Becker, Edward Yeary, and A. Gene Nelson
Jared M. Smalley
Edward Yeary and Manning Becker
James Lewis, Ed Glade, and Greg Gustafson
Ned D. Bayley

Part 2

William H. Pietsch
Daniel G. Piper
D. David Moyer

Part 3

Gerald E. Sherwood
Richard A. Biggs
James D. Wiseman
Theodore Brevik and Marion Longbotham
Fred Buscher and Jot Carpenter

E. F. Sedgley
Lee Allen
Stephen Berberich and Elmer Jones
Wesley Gunkel and David Ross

Part 4

Arthur Johnson
Nancy P. Weiss
John York and David Allan
Burl S. Ashley

Part 5

Harold Fogle and Miklos Faust
J.R. McGrew [grapes]
Gene Galletta, Arlen Draper, and Richard Funt [berries]
Allan Stoner and Norman Smith [vegetables]
Wesley Judkins and Floyd Smith [organic food]
 Martin Blum, Oray Stanton, and Raymond Williams
    [marketing produce]
Elizabeth Scholtz and Frederick McGourty Jr. [ornamental plants]
Francis Gouin and Ray Brush [plant nursery]
Dorothy Kuder Smith [dried flowers]
James A. Duke [herbs]
Conrad Link and David Ross [greenhouse gardening]
Norman E. Gary [beekeeping]
Maxwell L McCormack Jr. [Christmas trees]
Vernon Mayrose, James Foster, and Betty Drenkhahn [pigs]
R.A. Jaynes, G.C. Martin, L. Shreve, and G.S. Sibbett [nuts]
Hugh S. Johnson [chickens]
H. Travis, R. Aulerich, L. Ryland and J. Gorham [rabbits]
Kenneth G. MacDonald [beef cattle]
Larry Arehart [sheep]
Robert Appleman and Kenneth Thomas [dairy]
Donald L. Ace [goats]
Robert C. Church [horses]
Malcolm I. Bevins [vacation farms]

Part 6

J. W. Looney
Donald R. Levi

The editor also acknowledges the committee members who planned the original 1978 USDA Yearbook of Agriculture:

Ned D. Bayley, chairman
Alice Skelsey, deputy chairman
Jack Armstrong
William A. Bailey
Charles Beer
Richard A. Biggs
Wesley Harris
Pieter Hoekstra
Evelyn Johnson
James Lewis
Percy Luney
Charles McClurg
McKinley Mayes
David Ross
Floyd Smith
Billy Teels
Ralph Wilson
James D. Wiseman

Index

berries, 163–70
Bevins, Malcolm I., 316
Biggs, Richard A., 54–56, 315
biodiesel, 112
birds, as pests, 162
bittersweet, 192, 200, 201
blackberries, 163, 166
bleeding-heart, 191
blueberries, 165
Blum, Martin, 316
botrytis, 214
Boyce, Ed, 159
boysenberries, 166
bramble plants, 166
Brevik, Theodore, 315
browallia, 214
brucellosis, 265
Brush, Ray, 316
bugle, 204
building: barns, 96–98; codes and
    permits, 95; materials, 90–91;
    site, 90
buildings, farmstead, 89–100
bulbs, forcing of, 214
Buscher, Fred, 315

cabbage, 172
canned goods, 69
canning foods, 188
cantaloupes, 185
capons, 243, 244
carnations, 210
Carpenter, Jot, 315
carrots, 172
catnip, 205
cattails, 201
cattle: beef cattle, 258–65; dairy
    cows, 269–75
cauliflower, 172
cedar trees, 227
celosia, 201
chemicals, 21

cherries, 125, 150, 152, 182
chestnut blight fungus, 236–37
chestnut weevils, 237
chickens, 33, 97, 121; capons, 243; hen
    houses, 238–43; housing for, 98;
    waste as fertilizer, 123
children, supervision of, 182
China aster, 210
Chinese chestnut, 236–37
chives, 203
Christmas trees, 180, 182, 225–32
chrysanthemums, 210, 212
Church, Robert C., 316
clary, 205
climate, 31, 96, 121; building sites and,
    90; gardening and, 31; growth
    zones and, 126–27, 172; land
    purchase and, 83; strawberries,
    164–65; windows and, 48
coccidiosis, 257
cockscomb, 201
codling moth, 236
cold frames, 191, 205, 209
coleus, 204
collards, 172
Colony Collapse Disorder
    (bee disease), 223
columbine, 191
communities, alternative, 36–38
Community Supported Agriculture,
    184–85
compost, 22, 110, 127, 167, 176, 198
Concord grapes, 156, 158, 160
conservation, 87–88
container growing, 197–98
contracts, 297–300
Cooperative Extension offices, 19,
    21, 40, 52, 69, 87, 96, 307; berry
    culture and, 170; foresters and,
    140; 4–H program, 130, 307; fruit
    tree cultivation and, 154; green-
    houses and, 209; marketing

Gustafson, Greg, 315
harvesting, 154, 292; Christmas trees, 227, 231–32; grapes, 157; honey, 222–23; nuts, 235–36; pick-your-own, 169; vegetables, 173
hay, 127, 132; for cows and sheep, 125, 272; for horses, 100, 287–89
heating: greenhouse, 207, 208, 209–10; wood, 67–68
hens, 238–43
herbicides, 123, 175, 197, 211, 228, 237
herbs, 201, 203–6
hiking, 144, 291
hobby farmers, 10
homesteading, 36, 58
honey: bees, 215, 216; harvesting, 222–23; marketability, 122, 216, 219
horehound, 205
hornworms, 178
horses, 26, 85, 127, 282–83; breed selection, 283–85; feed require-ments, 287–88; horseback riding, 291; housing for, 100; pasture, 286; space and fencing for, 285–86; stables, 289–90
houses, 130–31; basement storage, 68–69; building, 57–63; earthen, 58; factory-built, 59; landscaping and, 75–76; paper-crete block, 58; remodeling, 45–53; rural residence, 31–32; straw bale, 58; sustainable, 58
humidity, 32, 50, 69, 213
hunting, 13, 143–44
hutches, for rabbits, 255
hyacinths, 214
hyssop, 204, 205

impatiens, 190, 214
income, 20, 22, 23
infectious bovine rhinotracheitis (rednose), 265

insect pests, 11, 121, 175–76, 177–78; of blueberries, 165; of Christmas trees, 227, 230; damage to houses, 49; in greenhouses, 211, 213; mints as repellents, 205–6; of nursery crops, 197; of nut trees, 236, 237; time of planting and, 177–78; tree damage and, 142–43
insulation, 49–50; inexpensive materials, 92–93
insurance, 95–96; for machinery, 115–16; rates, 15, 34; pick-your-own marketing, 182; stable, 282; for vacation farms, 293
intercropping, 205
irises, 191, 201, 214
irrigation, 20, 102; drip system, 84, 175; pasture improvement and, 83

Japanese beetles, 27, 176
Jaynes, R. A., 316
Johnson, Arthur T., 121–23, 124–28, 316
Johnson, Hugh S., 316
Johnston, John, 171
joint tenancy, 38, 304
Jones, Elmer, 316
Judkins, Wesley, 316
juice, 155

kale, 172
"killer" bees, 223–24

labor: berry culture, 169, 170; costs of, 114, 115, 116; dairy cows, 270, 271; greenhouse gardening, 208; sheep raising, 267; sources of, 20–21
land: buying, 10, 16–17; Christmas tree production, 225–27; cost of, 37; for dairy cows, 272–73;

improvements to, 81–88; investment in, 20; land use changes, 41–42; legal considerations, 39–40; transferring, 297–98

mowers, 116, 122
Moyer, D. David, 315
mulch, 168, 173, 175, 177, 178, 205, 213, 237

narcissus, 214
Natural Resource Conservation Service, 84
navel orange worm, 235
nectarines, 125, 149
neighbors, 13, 22, 27, 41, 90, 96, 238, 249, 290
Nelson, A. Gene, 315
nematodes, 149, 152, 156, 164, 165, 166, 168, 195, 235
Niagara grapes, 156
nurseries, 193–99
nut trees, 144, 233–37

oak trees, 196
O'Herron, Sara, 159, 160
okra, 172, 192, 201
orangemint, 204, 205
oregano, 203, 205
organic products, 176–77, 186–87
ornamentals, 190–92

paint, 47, 49, 50, 56, 143, 209, 210, 216
papercrete block house, 58
part-time farmers, 9, 17–18, 33–34
pasture: beef cattle and, 260–61; geese and, 245; for horses, 286; improvement of, 83; pigs and, 251
peaches, 125, 150, 152; pick-your-own, 182; sale to retail stores, 185
pears, 69, 125, 150, 154
peas, 130, 182, 201
pecan orchards, 233–34
peppermint, 203, 204, 205
peppers, 172, 173, 176
percolation tests, 41
perennials, 163, 165, 191, 196, 203, 204

pest control, 147, 151; alternative methods, 21; applicator's license and, 191; beekeeping and, 216, 223; chemical, 165; fruit, 165, 166, 167–68; fruit trees, 151; fungi, 164; for grapes, 157; for nut trees, 236; for vegetables, 175–78
petunias, 191, 210, 214
pH, of soil: berries, 164, 165, 166, 167; chestnut trees, 237; fruit trees, 152; mulch and, 176; poinsettias, 213; tomatoes, 213
pick-your-own, 181–82; berries, 149, 161, 163, 169, 173; grapes, 161; marketing, 181–82; vegetables, 172
Pierce's disease, 156
Pietsch, William H., 315
pigeons, 99
pigs, 97, 130, 248–49; feed needs, 250–51; shelter and equipment for, 98–99, 249–50; sow and newborns, 251–52; waste as fertilizer, 123
pine trees, 227
Piper, Daniel G., 315
plants: diseases of, 11; nurseries, 193–99
plumbing, 41, 51, 104, 105
poinsettias, 210, 212–13
pollination, 164, 166, 168, 215, 216
pollution: air, 13; water, 82, 85
ponds, 84–85, 106–7, 137, 138, 210
poppies, 202
potato beetles, 178
poultry, 238–47
privacy, 12, 13, 27, 52, 60, 76, 292
produce, marketing, 17, 121–22, 179–89
property, 297–300, 319–21
pruning, 141–42, 167, 196–97
public transportation, 34
pumpkins, 130, 172, 185